Springer Series in Reliability Engineering

For further volumes:
http://www.springer.com/series/6917

J. J. Xiong · R. A. Shenoi

Fatigue and Fracture Reliability Engineering

Springer

Prof. J. J. Xiong
Aircraft Department
Beihang University
Beijing
People's Republic of China
e-mail: jjxiong@buaa.edu.cn

Prof. R. A. Shenoi
School of Engineering Sciences
University of Southampton
Southampton
UK
e-mail: R.A.Shenoi@soton.ac.uk

ISSN 1614-7839

ISBN 978-0-85729-217-9 e-ISBN 978-0-85729-218-6

DOI 10.1007/978-0-85729-218-6

Springer London Dordrecht Heidelberg New York

British Library Cataloguing in Publication Data
A catalogue record for this book is available from the British Library

Cover design: eStudio Calamar, Berlin/Figueres

Printed on acid-free paper

Springer is part of Springer Science+Business Media (www.springer.com)

Preface

It has been reported that [1, 2] 80–90% of failures in load bearing structures are related to fatigue and fracture. Therefore, fatigue reliability analyses now are widely used to underpin design for safe operation of such artefacts. Fatigue loading on engineering structures results in the onset of damage which, from time to time, will require repair. This can be expensive if the structure/artefact has to be taken out of service for the repair to be effected. Occasionally, if the damage is not identified at an early stage, there is a likelihood of sudden, catastrophic failure. Thus it is important to determine, as precisely as possible, the service life and inspection periods in order to ensure safety. From practice, it is proved that because of the random nature of external loading on structure and the internal heterogeneity of the structural material and manufacturing variabilities, for the same style of structure under the same load conditions, the full-lives display large variations. Thus, it is difficult for a deterministic methodology to evaluate the service life of the product sample and to include the randomness above mentioned. Thus also there is a need for probabilistic approaches through a combination of probabilistic statistics and mechanics.

In order to guard against failures from unforseen circumstances, long-term efforts have been continually being put forward for enunciating newer and better approaches for imparting knowledge on reliability determination of fatigue and fracture behaviour, data treatment and generation of fatigue load spectrum, reliability design and assessment of structural total life, reliability prediction of composite damage and residual life, chaotic mechanism of fatigue damage, etc. through incorporating probability, statistics, stochastic process, non-linear random mathematics, fatigue, fracture mechanics and damage mechanics. Thus fatigue and fracture reliability engineering approaches to structural substantiation have been devised, which attempt is to decrease the structural failure probability resulting from fatigue or fracture to a lowest possible level for a structure to perform the given tasks under the given operation conditions during a given service period from economy viewpoint. The present book is an attempt to present an integrated and unified approach to related topics.

The importance of the subject has been recognised in recent years by many researchers and practitioners; it is taught in undergraduate and postgraduate programmes. A number of doctoral and research programmes are also being undertaken. Further, as part of the continuing education programme, many universities and commercial organisations are offering short term courses on this subject. A number of books already exist on the topic of fatigue and fracture reliability. They can broadly be classified under three headings. The first envisages the subject from a point of view of statistics, in which due to variations between individual specimens, fatigue data can be described by random variables to study the variability of fatigue damage and life and to analyze their average trends. Typical examples of this category are the works of Weibull [3], Freudenthal et al. [4] and Gao [5]. Books in the second category treat fatigue crack growth data as random fields/stochastic processes in a random time-space and state-space to depict local variations within a single specimen and to analyze the statistical nature of fatigue crack growth data. Examples are books by Bogdanoff and Kozin [6], Lin et al. [7], Provan [8] and Sobczyk and Spencer [9], etc. Books in third category deal with the reliability of structural components. Examples in this category are the works of Liard [10], etc.

This book transcends the traditional classifications mentioned above. Five distinguishing features of the new book are as follows.

1. A series of original and practical approaches including new techniques in determining fatigue and fracture performances, phenomenological expressions for generalized constant life curves, parameter estimation formulas, the two-dimensional probability distributions of generalized strength in ultra-long life regions are proposed. New techniques on randomization approach of deterministic equations and single-point likelihood method (SPLM) are presented to address the paucity of data in determining fatigue and fracture performances based on reliability concepts. Three new randomized models of time/state-dependent processes are presented for estimating the P-a-t, P–$da/dN-\Delta K$ and P–S–N curves, by using a randomization approach of deterministic equations and single-point likelihood method (SPLM), dealing with small sample numbers of data. The confidence level formulations for these curves are also given [11, 12]. Two new phenomenological expressions for generalized constant life curves are developed based on traditional fatigue constant life curve, and new parameter estimation formulas of generalized constant life curves are deduced from a linear correlation coefficient optimization approach. From the generalized constant life curves proposed, the original two-dimensional joint probability distributions of generalized strength are derived [13].
2. Novel convergence–divergence counting procedure is presented to extract all load cycles from a load history of divergence–convergence waves. The lowest number of load history sampling is established based on the damage-based prediction criterion. A parameter estimation formula is proposed for hypothesis testing of the load distribution [14]. An original load history generation approach is established for full-scale accelerated fatigue tests. Primary focus is

placed on the load cycle identification such as to minimize experimental time while having no significant effects on the new generated load history. The load cycles extracted from an original load history are identified into three kinds of cycles namely main, secondary and carrier cycles. Then the principles are presented to generate the load spectrum for accelerated tests, or a large percentage of small amplitude carrier cycles are deleted, a certain number of secondary cycles are merged, and the main cycle and the sequence between main and secondary cycles are maintained. The core of the generation approach is that explicit criteria for load cycle identification are established and equivalent damage calculation formulae are presented. These quantify the damage for accelerated fatigue tests [15].

3. Practical scatter factor formulae, dealing with conditions where the population standard deviation is unavailable and where fatigue test results are incomplete, are presented to determine the safe fatigue crack initiation and propagation lives from the results of a single full-scale test of a complete structure [16]. A new durability model incorporating safe life and damage tolerance design approaches is derived to assess the first inspection period for structures. New theoretical solutions are proposed to determine the s_a–s_m–N surfaces of fatigue crack initiation and propagation. Prediction techniques are then developed to establish the relationship equation between safe fatigue crack initiation and propagation lives with a specific reliability level using a two-stage fatigue damage cumulative rule [17].
4. The static and fatigue properties and the failure mechanisms of unnotched and notched CFR composite laminates with different lay-ups to optimize the stacking sequence effect are experimentally investigated, and it is seen that the process of composites fatigue damage under the compression cycles loading appears two different stages. The results of this study provide an insight into fatigue damage development in composites and constitute a fundamental basis for the development of residual strength model. Two new practical fatigue-driven models based on controlling fatigue stress and strain with four parameters are derived to evaluate fatigue residual strength easily and expediently from the small sample test data using the new formulae [18–21]. A dual cumulative damage rule to predict fatigue damage formation and propagation of notched composites is presented according to the traditional phenomenological fatigue methodology and a modern continuum damage mechanics theory. Then a three-dimensional damage constitutive equation for anisotropic composites is established. A new damage evolution equation and a damage propagation σ_a–σ_m–N^* surface are derived based on damage strain energy release rate criterion [22].
5. A nonlinear differential kinetic model is derived for describing dynamical behaviours of an atom at a fatigue crack tip using the Newton's second principle. Based on the theories of the Hopf bifurcation, global bifurcation and stochastic bifurcation, the extent and some possible implications of the existence of atomic-scale chaotic and stochastic bifurcative motions involving the fracture behaviour of actual materials are systematically and qualitatively

discussed and the extreme sensitivity of chaotic motions to minute changes in initial conditions is explored. Chaotic behaviour may be observed in the case of a larger amplitude of the driving force and a smaller damping constant. The white noise introduced in the atomistic motion process may lead to a drift of the divergence point of the non-linear stochastic differential kinetic system in contrast to the homoclinic divergence of the non-linear deterministic differential kinetic system [23]. By using the randomization of deterministic fatigue damage equation, the stochastic differential equation and the Fokker–Planck equation of fatigue damage affected by random fluctuation are derived. By means of the solution of equation, the probability distributions of fatigue crack formation and propagation with time are obtained [24].

To the best of the authors' knowledge, no book on fatigue and fracture reliability engineering has been written so far based on the above considerations.

The book is intended for practising engineers in marine, civil construction, aerospace, offshore, automotive and chemical industries. It should also form a useful first reading for researchers on doctoral programmes. Finally, it will also be appropriate for advanced undergraduate and postgraduate programmes in any mechanically-oriented engineering discipline.

August 30, 2010

J. J. Xiong
R. A. Shenoi

References

1. Committee on fatigue and fracture reliability of the Committee on structural safety and reliability of the structural division (1982) Fatigue reliability 1–4, Journal of Structural Division, Proceedings of ASCE 108 ST1:3–88
2. Cheung MMS, Li W (2003) Probabilistic fatigue and fracture analysis of steel bridges. J Structural Safety 23:245–262
3. Weibull W (1961) Fatigue testing and analysis of results. Macmillan Company, New York
4. Freudenthal AM, Garrelts M, Shinozuka M (1966) The analysis of structural safety. J Struct Div, ASCE 92:267–325
5. Gao ZT (1981) Applied statistics in fatigue. National Defense Press, Beijing
6. Bogdanoff JL, Kozin F (1985) Probabilistic models of cumulative damage. Wiley, New York
7. Lin YK, Wu WF, Yang JN (1985) Stochastic modeling of fatigue crack propagation: probabilistic methods in mechanics of solids and structure. Springer, Berlin
8. Proven JW (1987) Probabilistic fracture mechanics and reliability. Martinus Nijhoff, Dordrecht (The Netherlands)

9. Sobczyk K Jr, Spencer BF (1992) Random fatigue-from data to theory. Academic Press, Inc, London
10. Liard F (1983) Helicopter fatigue design guide. AGARD-AG-292
11. Xiong JJ, Shenoi RA (2007) A practical randomization approach of deterministic equation to determine probabilistic fatigue and fracture behaviours based on small experimental data sets. Int J Fracture 145:273–283
12. Xiong JJ, Shenoi RA. (2006). Single-point likelihood method to determine a generalized S–N Surface. Proceedings of the I Mech E (Institution of Mechanical Engineers) Part C J Mech Eng Sci 220(10):1519–1529
13. Xiong JJ, Shenoi RA, Zhang Y (2008) Effect of the mean strength on the endurance limit or threshold value of the crack growth curve and two-dimensional joint probability distribution. J Strain Anal Eng Des 43(4): 243–257
14. Xiong JJ, Shenoi RA (2005) An integrated and practical reliability-based data treatment system for actual load history. Fatigue Fract Eng Mater Str 28(10): 875–889
15. Xiong JJ, Shenoi RA (2008) A load history generation approach for full-scale accelerated fatigue tests. Eng Fract Mech 75(10):3226–3243
16. Xiong J, Shenoi RA, Gao Z (2002) Small sample theory for reliability design. J Strain Anal Eng Des 37(1):87–92
17. Xiong JJ, Shenoi RA (2009) A Durability model incorporating safe life methodology and damage tolerance approach to assess first inspection and maintenance period for structures. Reliab Eng Syst Saf 94:1251–1258
18. Xiong JJ, Shenoi RA, Wang SP, Wang WB (2004) On static and fatigue strength determination of carbon fibre/epoxy composites. Part II: Theoretical formulation. J Strain Anal Eng Des 39(5):541–548
19. Xiong JJ, Shenoi RA, Wang SP, Wang WB (2004) On static and fatigue strength determination of carbon fibre/epoxy composites. Part I: Experiments. J Strain Anal Eng Des 39(5):529–540
20. Xiong JJ, Li YY, Zeng BY (2008) A strain-based residual strength model of carbon fibre/epoxy composites based on CAI and fatigue residual strength concepts. Composite Struct 85:29–42
21. Xiong JJ, Shenoi RA (2004) Two new practical models for estimating reliability-based fatigue strength of composites. J Composite Mater 38(14):1187–1209
22. Xiong JJ, Shenoi RA (2004) A two-stage theory on fatigue damage and life prediction of composites. Composites Sci Tech 64(9):1331–1343
23. Jun-Jiang Xiong (2006) A nonlinear fracture differential kinetic model to depict chaotic atom motions at a fatigue crack tip based on the differentiable manifold methodology. Chaos Solitons Fractals 29(5):1240–1255
24. Xiong JJ, Gao ZT (1997) The probability distribution of fatigue damage and the statistical moment of fatigue life. Sci China Ser E 40(3):279–284

Contents

Chapter 1
Deterministic Theorem on Fatigue and Fracture

1.1 Introduction

In the mid-1800s, with the appearance and development of the railways, after a number of loading cycles, a likelihood of sudden, catastrophic failure usually appeared on the axle shafts of locomotives. The phenomenon had received considerable attention. From observation, it was found that failure in the axle shaft resulted from a cyclic bending stress that was less than the static endurance limit. Since 1849, fatigue experiments on the axle shafts of locomotives under a cyclic load regime revealed the relationship between the cyclic stress and number of load cycles to fatigue failure. It was observed that the number of load cycles to fatigue rupture decreased with the increasing cyclic stress. Further, when the cyclic stress was less than a specific value, no fatigue failure would occur regardless of the numbers of load cycles. The relationship between the cyclic stress and number of load cycles was termed as the S–N curve. With the expansion of the railways, more and more fatigue failures appeared, which resulted in further studies and researches on fatigue behaviour.

In likewise manner, aircraft structures in service are also subject to cyclic loading which, in turn, leads to onset of fatigue damage. Early approaches to aircraft design relied principally on static strength limits albeit with high enough safety factors to inhibit the inducement of fatigue damage. However, with the rapid development of modern aeronautics, it was necessary for the weight of aircraft to be as light as possible for high speeds and enhanced flight performance. This required more accurate determination of the static strength and a better understanding of safety margins in service. Specifically, owing to the longer service times and with working stress levels being nearer failure limits, fatigue problems in civil aircrafts became more severe, leading to some accidents such as the Comet disaster in 1954. These catastrophic accidents received extensive attention and a large number of theoretical and experimental investigations have since been carried out to understand the mechanism and laws on fatigue for

J. J. Xiong and R. A. Shenoi, *Fatigue and Fracture Reliability Engineering*,
Springer Series in Reliability Engineering, DOI: 10.1007/978-0-85729-218-6_1,

aircraft. The resultant improvements in modeling have been applied widely. The concepts behind such modeling are introduced below.

1.2 Fatigue Failure Character and Fracture Analysis

As mentioned above, a failure phenomenon under cyclic loading is termed as fatigue. The main characteristics of fatigue failure are as follows:

1. Failure occurs at a stress level less than the static strength of the material or structure.
2. Fatigue always proceeds some time, even long time.
3. For plastic materials, in general, there is no significant permanent deformation before fatigue failure. Therefore, it is not easy for fatigue damage to be detected during inspection and maintenance procedures.
4. The fractographs of fatigue failure (illustrated in Figs. 1.1 and 1.2) show smooth and coarse zones. After a certain number of load cycles, a micro-crack first appears at a site termed as the fatigue origin. From the fatigue origin, the crack gradually propagates away, extruding and separating to induce a smooth zone. With the crack propagation, the section subjected to the loading is weakened severely until it is unable to resist the loading and a sudden rupture takes place.

Some other observations based on failure studies are as follows. In contrast to brittle materials, there is no significant permanent deformation in plastic material before fatigue failure. Consequently, it is argued that failure owing to fatigue is the result of a change of internal structure of the material under cyclic loading, or in other words, the "fibre" structure of a plastic material degenerated into the "crystals" structure of a brittle material. It is further reported that fatigue failure consisted of three stages as: (1) fatigue crack formation (initiation), (2) stable

Fig. 1.1 Fatigue fractograph diagram

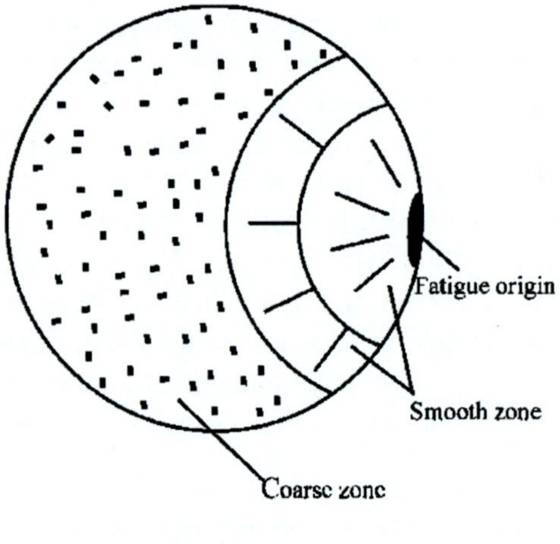

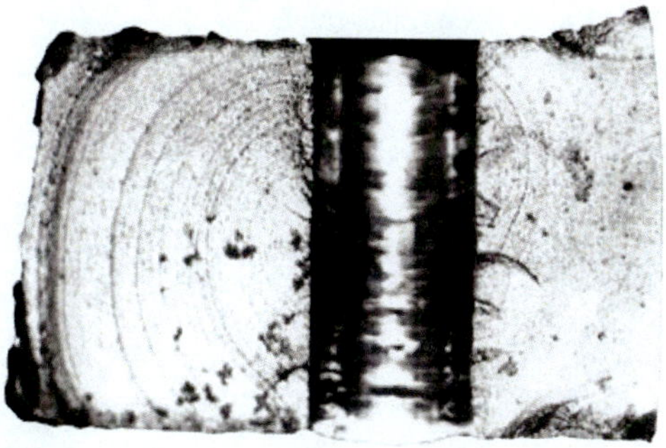

Fig. 1.2 Fatigue fractograph

Fig. 1.3 Fatigue fractograph of blade of engine

fatigue crack propagation and (3) unstable fatigue crack propagation to result in a sudden breakdown.

During the tensile tests of low carbon steel, if the tensile stress exceeds yield limit, then fracture surfaces of specimen exist minute and close slip lines whose direction has an angle of 45° with the axis of specimen. It is interesting that under cyclic stress, there are the same slip lines on particular zones of fracture surface of specimen which become thicker with increasing load cycles. In the case that fatigue stress is greater than the fatigue limit, the slip lines and the striation spacings between two adjacent and sequential slip lines become rougher and larger; there are slip lines on fracture surfaces of electropolished specimen too. These remaining slip lines are termed as the resident slip lines. In fact, the resident slip lines are the micro cracks and would become into the rougher slip bands with the increasing of loading cycles. Fatigue origins generally emerge at the zone with the most compact slip lines among resident slip bands.

As mentioned above, it is known that the conchoidal marking lines usually occur on smooth zones of fracture surfaces, which can be seen through naked eyes and are known as the fatigue lines, or leading-edge lines or resting lines. Fatigue rupture surfaces of the blade of an engine and of the lower lug of wing spar are shown in Figs. 1.3 and 1.4 respectively. Figures 1.3 and 1.4 show clear fatigue lines for interpreting fracture surfaces.

After the formation of an initial fatigue crack, fatigue crack continues to propagate with increasing load cycles. In the case of constant amplitude of cyclic stress and homogeneous material, during stable propagation of fatigue crack no conchoidal marking line occurs on fracture surfaces. Fatigue lines appear very rarely during tests of standard coupon subjected to cyclic stress with constant amplitude. However, actual structures in service are subjected to cyclic stress with variable amplitude and the structural materials generally are inhomogeneous. As a result, in case of the changes in stress amplitude, crack growth rate varies and this leads to a circumferential fatigue line on fracture surfaces, which is at the crack

Fig. 1.4 Fatigue fractograph of lower lug of wing spar

leading-edge and hence also termed as the loading-edge line. Obviously, the variability of stress amplitude and the inhomogeneity of material are important reasons to cause conchoidal marking lines on fracture surfaces. As is shown in Figs. 1.1 and 1.2, there are some segmental radial lines between circumferential fatigue lines among conchoidal marking lines, which result from a sudden and large loading. In general, a sudden and large loading leads to the propagations of crack along different plane surfaces to misalign a segmental radial line between different plane surfaces, therefore, it stands to reason that the radial lines occur. Circumferential fatigue lines might also appear along the edge between adjacent smooth and coarse zones of a homogeneous material subjected to cyclic stress with constant amplitude; this implies that a more rapid propagation of fatigue crack takes place in short time to cause the fatigue line.

1.3 Cyclic Stress and S–N Curve

Cyclic stress varies periodically and is denoted as s. It is worth pointing out that in order to distinguish both physical quantities of fatigue stress and strength, the former is denoted as s, whereas the latter as S. Ordinarily, cyclic stress is assumed to vary in a sinusoidal form, and hence implemented for depicting the change of cyclic stress with time (as shown in Fig. 1.5). Within one stress cycle, the highest value is called as the maximum stress s_{max} and the lowest is termed as the minimum stress s_{min}. The algebraic mean of maximum and minimum stresses is defined as the mean stress s_m:

$$s_m = \frac{s_{max} + s_{min}}{2} \tag{1.1}$$

As shown in Fig. 1.5, s_a is known as the stress amplitude:

$$s_a = \frac{s_{max} - s_{min}}{2} \tag{1.2}$$

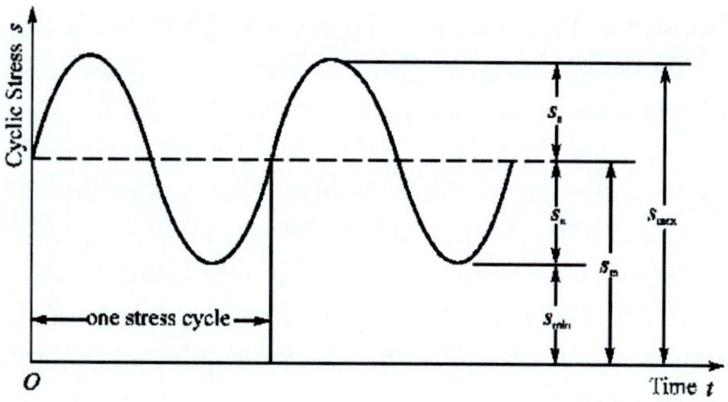

Fig. 1.5 Cyclic stress

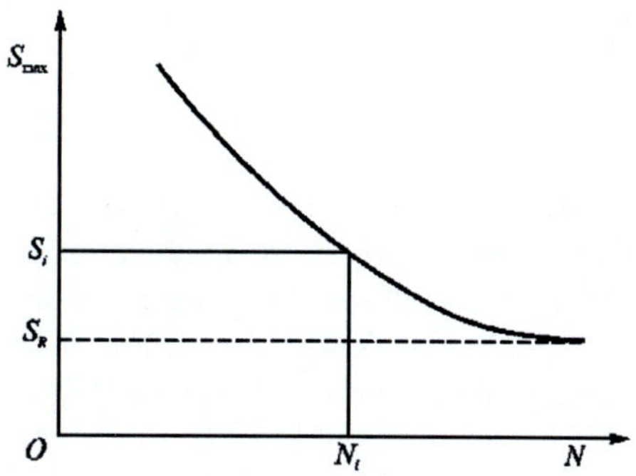

Fig. 1.6 S–N curve of conventional steel and cast-iron

The ratio of minimum stress to maximum stress is defined as the stress ratio:

$$R = \frac{s_{\min}}{s_{\max}} \tag{1.3}$$

From Fig. 1.5, it is seen that a cyclic stress could be conceived of comprising two components, namely a static stress of magnitude the same value as the mean stress s_{m}, and a dynamic stress component that varies symmetrically around the mean stress. Five physical quantities of a cyclic stress including $s_{\max}$, s_{m}, s_{m}, s_{a}, R have three relationships expressed by using Eqs. 1.1–1.3; therefore only two physical quantities are independent and adequate to describe a cyclic stress, e.g., $s_{\max}$ and R, or s_{a} and s_{m}, whereas static stress is depicted only through one quantity (i.e., the stress value).

Under cyclic loading, the resistance of the material to fatigue is generally represented using the S–N curve and a fatigue limit. In the case of a specific stress ratio of R, a set of standard coupons are subjected to cyclic loading with different maximum stress $S_{\max}$ until failure to obtain the cycle number N of each specimen to fatigue failure. Thus the curve with the vertical axis of $S_{\max}$ and horizontal axis of cycle number N to failure is the S–N curve of material at the specific stress ratio of R. Obviously, a pair of quantities of (S, N) are required to depict fatigue strength of material.

Experiments show that the S–N curves of conventional steel and cast-iron, etc. exhibit a horizontal asymptotic line (shown in Fig. 1.6) as: $S_{\max} = S_R$. This implicates that if $S_{max} > S_R$, the specimen then fails after a certain cycle number, while if $S_{max} < S_R$, no failure of specimen occurs. S_R is known as the fatigue limit of material at a specific stress ratio, R. In the case of $R = -1$ (i.e., complete reversal loading), fatigue limit is denoted as S_{-1}. However, there is no horizontal asymptotic line on the S–N curves for nonferrous metals and their alloys, thereby the S_{max} relevant to a specific cycle number of N (e.g., 10^7 cycles) to failure is generally regarded as the conditional fatigue limit.

Many works have been performed to expose the laws of the S–N curve. Due to the large diversity of S–N curves for different materials, it is difficult to establish a unified expression of an S–N curve. It is generally accepted that S–N representation can be done using four empirical formulas, as follows.

(1) Power function expression:

$$S_a^m N = C \tag{1.4}$$

where m and C are the material constants associated with material property, specimen configuration, loading mode, etc. and determined from experiments. Equation 1.4 shows the power function relationship between stress amplitude S_a and life N at a specific stress ratio of R or mean stress of S_m and is also applied to express the relationship between S_{max} and N, namely

$$S_{max}^m N = C \tag{1.5}$$

(2) Exponent function expression:

$$e^{mS_{max}} N = C \tag{1.6}$$

where e is the base of a natural logarithm. m and C are the material constants determined from experiments. Equation 1.6 represents the exponential function relationship between maximum stress S_{max} and life N under a specific stress ratio of R or mean stress of S_m.

(3) Three-parameter power function formula:

$$(S_{max} - S_0)^m N = C \tag{1.7}$$

or, alternatively

$$S_{max} = S_\infty \left(1 + \frac{A}{N^\alpha}\right) \tag{1.8}$$

where S_0, m, C, A, α and S_∞ are the material constants, and $C = (A \cdot S_\infty)^{1/\alpha}$, $m = 1/1\alpha$, $S_0 = S_\infty$. S_0 and S_∞ are the fatigue strength S_{max} in the case of $N \to \infty$, approximate the fatigue limit.

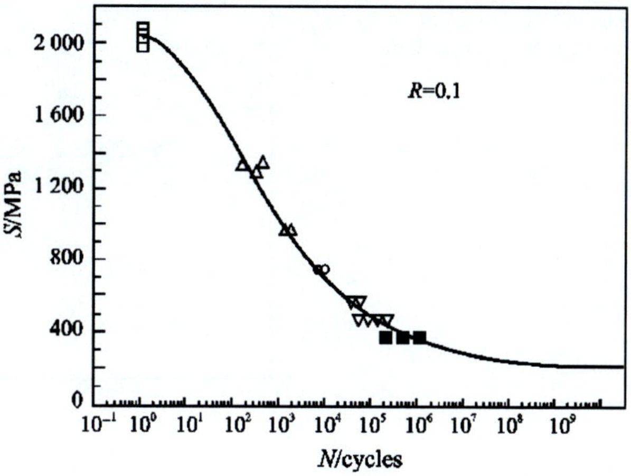

Fig. 1.7 Full-range S–N curve with four parameters

(4) Four-parameter power function formula:

$$\left(\frac{S_u - S_0}{S - S_0}\right) = 10^{C(\log N)^m} \tag{1.9}$$

where S_0, m, C and S_u are the material constants. m is the shape parameter. S_0 is the fitted fatigue limit. S_u is the fitted yield limit. Equation 1.9 has following physical features as (shown in Fig. 1.7): in the case of $N = 1$, then $S = S_u$; whereas in the case of $N = \infty$, then $S = S_0$. Fatigue life N varies with fatigue stress S, where for greater S one observes less N while for lower S, one observes larger N.

Noting that Eqs. 1.4–1.6 are valid only for expressing medium life range of S–N curve. Equations 1.7 and 1.8 are appropriate for medium and long life range of S–N curve and have a better fitting accuracy than those of Eqs. 1.4–1.6 because of three undetermined parameters. Equation 1.9 covers the full life range of S–N curve and the fitting precision is the best owing to the four characterising parameters.

1.4 Constant Life Curve and Generalized Fatigue S–N Surface

As mentioned above, for a specific stress ratio, an S–N curve of material can be determined from a set of fatigue experimental data. Since the S–N curve of material varies with the stress ratio R, this means a group of S–N curves of material can be obtained corresponding to different stress ratios R. According to the stress ratios R and relative fatigue strengths S_{max}, it is possible to calculate S_a and S_m and a constant life curve can be drawn in the coordinate system with the ordinate being S_{max} or S_a and the abscissa being S_m (shown in Fig. 1.8). Based on the statistical analysis of fatigue experimental data, several empirical equations are proposed to depict the constant life curve (shown in Fig. 1.8) as follows:

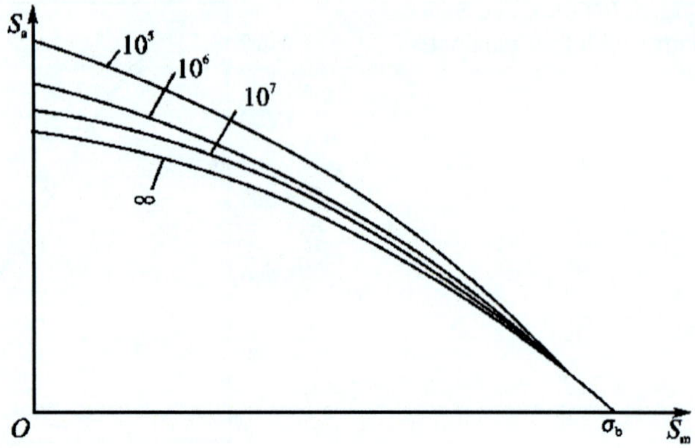

Fig. 1.8 Constant life curve relevant to different fatigue lives

(1) Goodman equation [1]:

$$\frac{S_a}{S_{-1}} + \frac{S_m}{\sigma_b} = 1 \tag{1.10}$$

where σ_b is the tensile ultimate strength of material. S_{-1} is the fatigue limit of material under complete reversal loading.

(2) Gerber parabola equation [2]:

$$\frac{S_a}{S_{-1}} + \left(\frac{S_m}{\sigma_b}\right)^2 = 1 \tag{1.11}$$

(3) Soderberg equation [3]:

$$\frac{S_a}{S_{-1}} + \frac{S_m}{\sigma_s} = 1 \tag{1.12}$$

where σ_s is the tensile yield limit of material.

(4) Shieliasan folding line [4]:

$$\frac{S_a}{S_{-1}} + \left[\left(\frac{2}{S_0} - \frac{1}{S_{-1}}\right)S_m\right] = 1 \tag{1.13}$$

where S_0 is the fatigue limit of material under pulsation cycle loading.

(5) Bagci quartic equation [5]:

$$\frac{S_a}{S_{-1}} + \left(\frac{S_m}{\sigma_s}\right)^4 = 1 \tag{1.14}$$

By re-examining the models (1.10)–(1.14), it is understood that [4] Eq. 1.12 is conservative for most of materials. Equation 1.14 is optimistic for most of materials. Equation 1.10 is adopted for brittle materials while being conservative for ductile materials. Equation (1.11) is suitable for ductile materials, but its use is limited because of its non-linearity function. Equation (1.13) can fit the test data perfectly, but it must be used at the given material fatigue endurance limit under the pulsating cyclic load. It is also found that these five kinds of constant life curves are unfit for most materials because the power exponents of these materials do not accord with these constant life curves. Some are between 1 and 2, and others are between 2 and 4. Hence, for the different materials, the power exponents of fatigue constant life curve are also different. From the literature [6], it is clear that many models are dependent on several undetermined reference variables. Further, the non-linear and exponential nature makes them computationally complex. The accurate and logical method for estimation of parameters has not been given. Therefore, based on the above models, a new and phenomenological generalized fatigue constant life curve formula for all kinds of material is proposed [7]:

$$\frac{S_a}{S_{-1}} + \left(\frac{S_m}{\sigma_b}\right)^m = 1 \tag{1.15}$$

where S_{-1}, σ_b and m are the material constants determined from fatigue experiments. σ_b is the fitting tensile ultimate strength. S_{-1} is the fitting fatigue limit under complete reversal loading. Equation 1.15 is termed as the generalized fatigue constant life curve.

The high cycle fatigue performance characterized by a S–N curve is determined at a given mean stress S_m or stress ratio R (namely, S_{min}/S_{max}, where S_{min} is the minimum stress and S_{max} is the maximum stress). In case of $S_m = 0$, or ($R = -1$), S_{max} represents symmetric cyclic fatigue strength S_{-1}, and thus, Eq. 1.7 can be written as:

$$(S_{-1} - S_0)^m N = C \tag{1.16}$$

From Eq. 1.10, the Goodman constant life line can be expressed as:

$$S_a = S_{-1}\left(1 - \frac{S_m}{\sigma_b}\right) \tag{1.17}$$

From Eqs. 1.16 and 1.17, it is possible to obtain:

$$\left(\frac{\sigma_b}{\sigma_b - S_m} S_a - S_0\right)^m N = C \tag{1.18}$$

where C, m and S_0 are the undetermined parameters. It is seen that Eq. 1.18 describes the relationship between stress amplitude S_a, mean stress S_m and fatigue life N, displaying an S_a–S_m–N surface in three-dimensional coordinate system.

1.5 Stress State and Growth Mode of Penetrated Crack

The fatigue process includes the two-stages of crack initiation and crack propagation and there is a considerable long time from crack initiation to fracture, the total fatigue life of a structural component is then equal to the sum of the time to crack initiation (i.e., to visually detectable crack size) and the time for the crack to propagate until it reaches a critical size. Sometimes, premature accidental failure of some important load-carrying structural components in service took place even though these may have been designed using an S–N curve, albeit of smooth specimens. Defects and imperfections in manufacture and service, such as forging defects, weld induced inclusions, surface scratch marks, corrosion pits, etc., usually led to surface cracks or inherent defects in structural components which would propagate until a sudden facture under cyclic loading. As a result, it was necessary to accept the existence of structural crack as a fact and to recognize the crack growth properties under cyclic loading in order to improve and complete fatigue design.

With the rapid development of manufacturing techniques for airplanes, rockets, ships, etc., a large number of brittle facture accidents arising from fatigue failure induced stricter and stricter requirements for structural design. A fail-safe design was introduced to complete safe-life design, in which the damage would be temporarily tolerated until repair could be effected, failing which the damage could assume potentially critical dimensions Thus it is necessary to investigate crack growth rate and resistance capability of material.

According to fracture theorem, fracture strength criterion is:

$$\sigma < \sigma_b$$

where σ_b is the ultimate strength of material and the primary index of facture of material. From the above equation, it is apparent that greater the value of σ_b, stronger is the resistance capability of material to fracture. Therefore, in engineering practice of aircraft and ship design, a higher strength of material was required to obtain greater load-carrying capability at a lighter weight. As a result of a large amount of research to improve strength limit σ_b of material for a certain long time, the improvement of smelting technique and rapid development of new materials and new processing constantly caused a large number of materials with higher strength, even greater than 2000 MPa.

Extensive applications of material with high strength led to a new problem, i.e., continually catastrophic fracture accidents under lower stress in engineering practice. In other words, even in case where the operating stress was lower than strength limit σ_b of a material, the structural component still ruptured. From these catastrophic fracture accidents, it has been found that the materials with high strength σ_b sometimes are not strong to resist against fracture.

Fracture can be categorized under three categories of opening, shearing and tearing modes, which are termed as the modes I, II and III respectively (shown in

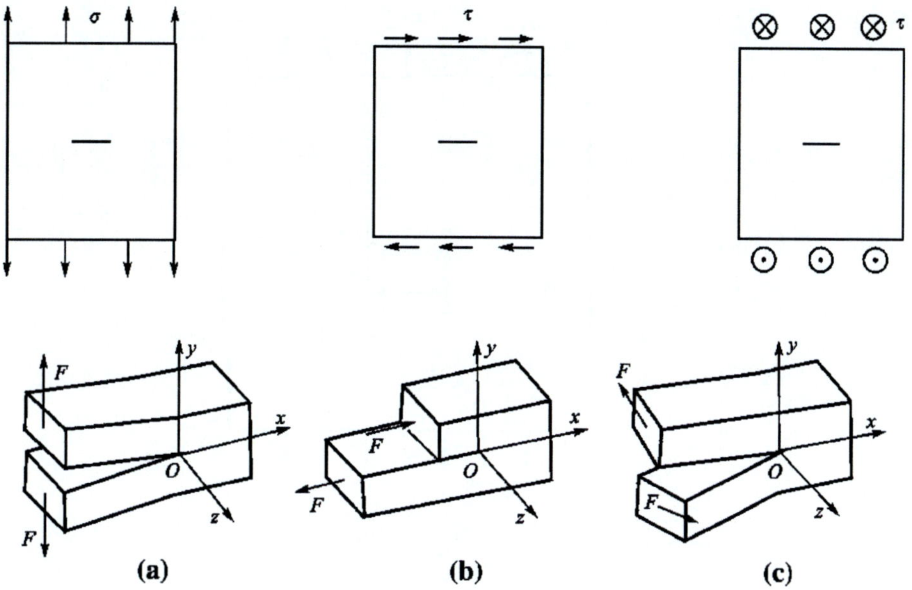

Fig. 1.9 Three categories of basic crack modes

Fig. 1.9). Figure 1.9 shows a penetrated crack from the upper surface to lower surface.

1. Under tensile stress σ perpendicular to the crack plane surface, the crack is opened; this is termed as the opening mode, mode I.
2. Under shear stress τ parallel to the crack plane surface and perpendicular to the crack tip, the crack is staggered along the crack plane surface; this is termed as the shearing mode, mode II.
3. Under shear stress τ parallel to both the crack plane surface and the crack tip, the crack tears the material; this is termed as the tearing mode, mode III.

Mode I (i.e., opening mode of loading) is the most frequent and hazardous for inducing the brittle fracture, thus the most investigations have been focused on this mode of loading. In general, the plastic zone near a crack tip is regarded as very small and hence linear elastic mechanics can then be used to analyse the crack behaviour. Both stress and strain fields near a crack tip may be characteries only through a parameter of K, called as the stress intensity factor.

A notched plate with a penetrated crack of $2a$ length perpendicular to the uniform tensile stress of σ, with a being much less than the length and width of plate and with the upper and lower end-edges being far from the penetrated crack (shown in Fig. 1.10), is regarded as an infinite plate. If the plate has enough thickness, then the plate is regarded as being in a the plane strain state.

According to linear elastic mechanics, the stress components σ_x, σ_y, τ_{xy} at any point (its polar coordinate (ρ, ϕ) as shown in Fig. 1.10) are respectively:

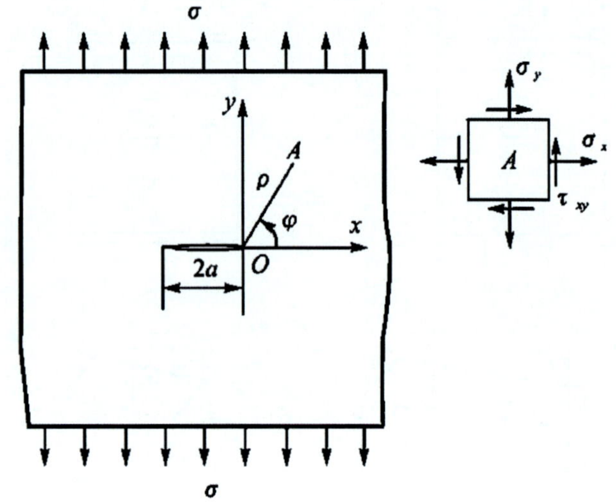

Fig. 1.10 Stress field near crack tip

$$\begin{aligned} \sigma_x &= \frac{K_{\mathrm{I}}}{\sqrt{2\pi\rho}} \cos\frac{\phi}{2}\left[1 - \sin\frac{\phi}{2}\sin\frac{3\phi}{2}\right] \\ \sigma_y &= \frac{K_{\mathrm{I}}}{\sqrt{2\pi\rho}} \cos\frac{\phi}{2}\left[1 + \sin\frac{\phi}{2}\sin\frac{3\phi}{2}\right] \\ \tau_{xy} &= \frac{K_{\mathrm{I}}}{\sqrt{2\pi\rho}} \sin\frac{\phi}{2}\cos\frac{\phi}{2}\cos\frac{3\phi}{2} \end{aligned} \tag{1.19}$$

With:

$$K_{\mathrm{I}} = \sigma\sqrt{\pi a} \tag{1.20}$$

Equation 1.19 ignores the higher order items of ρ and may be employed only for a very small area near the crack tip, or in the area where ρ is much less than the crack length. Equation 1.19 reveals the stress distribution near the crack tip, from which the stress σ_y at the point near the crack tip with a distance of ρ from crack tip O on the extension line of crack can be determined as:

$$\sigma_y = \sqrt{\frac{a}{2\rho}} \cdot \sigma \tag{1.21}$$

From Eqs. 1.19 and 1.21, it is seen that the relative strength of stress field near crack tip is governed only by a parameter of K_{I}, the stress intensity factor. From Eq. 1.19, it is also observed that in case $\rho \to 0$, the stress component tends to infinity. Actually, it is impossible for the stress component near crack tip to reach infinity since a small plastic zone near crack tip would result from the yielding of the material. Then it is difficult to utilize directly the stress quantity near crack tip as the criterion of unstable crack growth.

The limiting value of the stress intensity factor of a material is given as:

$$K_{\mathrm{I}} = K_{\mathrm{IC}} \tag{1.22}$$

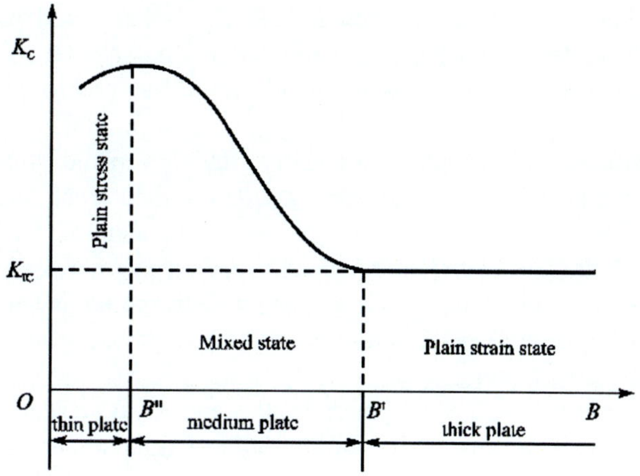

Fig. 1.11 Relationship curve between K_C and thickness B

where K_{IC} is the plane strain fracture toughness, representing the resistance of material to crack growth under a three-dimensional tensile stress state.

It is worth emphasizing that from a physical purpose, K_I is a parameter to depict the stress and strain fields near a crack tip and represents the service state of a notched structural component, whereas K_{IC} is representative of the inherent nature of material. From Eq. 1.22, it is clear that under general environmental and static loading, it is necessary to control the working stress intensity factor K_I below the critical value K_{IC} for the safe service of a notched structural component.

From elastic mechanics, the stress formulations under both plane stress and strain states have same forms, and the stress σ_y at the point near the crack tip with a distance of ρ from crack tip O on the extension line of crack is:

$$\sigma_y = \sqrt{\frac{a}{2\rho}} \cdot \sigma \tag{1.23}$$

For an infinite plate under plane stress state, stress intensity factor of mode I becomes:

$$\bar{K}_I = \sqrt{\pi a} \cdot \sigma \tag{1.24}$$

Thus, using Eq. 1.22 as an analogy, the fracture strength criterion for a notched component under plane stress state can be written as:

$$\bar{K}_I < K_C \tag{1.25}$$

where K_C is the fracture toughness of material under plane stress conditions. It is worth pointing out that although both K_{IC} and K_C represent fracture toughness, K_{IC} is a constant and a basic parameter of material, while K_C varies with the thickness of a plate. Figure 1.11 demonstrates the relationship between the fracture toughness K_C and thickness B. From Fig. 1.11, it is apparent that under a plane stress state, K_C is variable, and tends to a constant after B reaches a specific thickness of B', i.e., $K_C \rightarrow K_{IC}$ (a constant) and the stress state tends to the plane strain state.

The stress state between $B^{'}$ and $B^{''}$ is the mixed plane stress and strain state. Therefore, under plane strain state, by calculating and comparing the stress intensity factor K_{I} with plane strain fracture toughness K_{IC} determined from fracture experiments of thick plates, it is possible to predict the fracture limits of notched components. Similarly, under a plane stress state, it is necessary to compare fracture toughness K_{C} determined from fracture experiments on specimens the same thickness as the component under consideration. Though the fracture problem under a plane stress state is more complex than that under a plane strain state, thin plates have been widely applied in aeronautics, ship building, steel bridges, etc. and thus it is essential for fracture problem under plane stress state to be investigated.

Both fracture toughness of K_{IC} and K_{C} represent facture strengths of material under static loading condition, while fatigue crack growth threshold ΔK_{th} denotes facture strength of material subjected to cyclic loading and is defined as the lower limit value of ΔK to induce fatigue crack growth. In other words, fatigue crack would propagate at a greater ΔK and the growth rate would decrease with decreasing ΔK. When ΔK decreases to a specific value of ΔK_{th} then the fatigue crack stops to propagate. It is worth noting that with decreasing ΔK the plastic zone near crack tip becomes smaller and crack tip becomes sharper. As a consequence, it stands to reason that ΔK_{th} is independent on the size of plastic zone near crack tip.

ΔK_{th} is dependent not only on material's state but also on stress state (plane stress or strain). Moreover, ΔK_{th} depends on environmental factors (temperature, humidity, etc.). ΔK_{th} is also a function of cyclic stress ratio R. However, experimental observations reveal that under same material state, thickness, environmental condition and stress ratio, if crack length and ligament size satisfy linear elastic condition, then ΔK_{th} is independent of the configuration and dimension of specimen and becomes a material constant.

Since it is hard to absolutely stop crack growth to determine ΔK_{th}, it is necessary to approximate the condition of stoping crack propagation (or $\mathrm{d}a/\mathrm{d}N = 0$). That is, ΔK_{th} is defined as the value of ΔK pertaining to the crack growth rate of $\mathrm{d}a/\mathrm{d}N = 10^{-7}$mm/cycle; in other words, $\mathrm{d}a/\mathrm{d}N = 10^{-7}$mm/cycle is employed to approximate $\mathrm{d}a/\mathrm{d}N = 0$. As an exemplar, $\Delta K^{'}_{\mathrm{th}}$ as shown in Fig. 1.12 is used to approximate ΔK_{th}.

Fatigue crack growth threshold ΔK_{th} plays an important role in structural damage tolerance design. For a notched component subjected to a cyclic loading in service, it is effective and conservative to control the working stress intensity factor ΔK to remain below ΔK_{th} of the material under the service condition of the component in order to assure its safe operation. Additionally, ΔK_{th} can also be applied as the criterion to delete small load cycles while generating a fatigue load spectrum for crack growth tests and used in generalized Wheeler [8] and Willenberg et al. [9] models for crack growth analysis considering the overload retarding effect. The methods to determine fatigue crack growth threshold include (1) continuously decreasing load (or decreasing K) procedure, (2) percentage

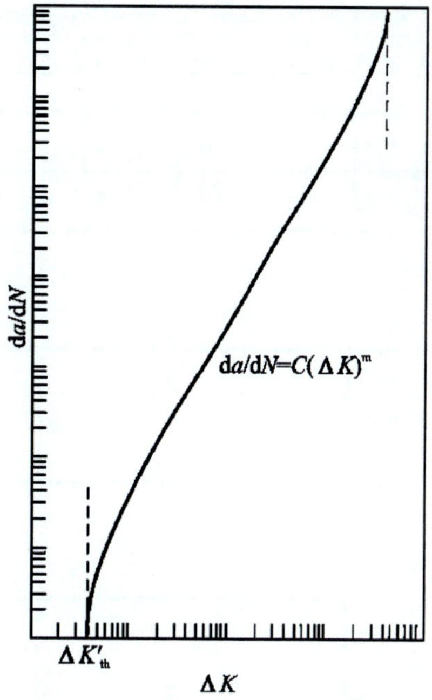

Fig. 1.12 Definition of ΔK_{th}

step-decreasing load process, (3) gradient K method, controlling constant P and (4) gradient K method, controlling constant K, etc.

The above formulations on stress intensity factor and fracture toughness are valid and effective for the infinite plane plates (or $W >> 2a$). In reality, the plates always have a finite width and sometimes the crack length reaches a fraction of plate's width. Then Eq. 1.20 needs to be modified by a correction factor α (shown in Table 1.1) to become applicable. Thus, the formula of stress intensity factor corresponding to mode I with a central crack of finite wide plate can be written as:

$$K_I = \sqrt{\pi a} \cdot \sigma \cdot \alpha \quad (1.26)$$

As mentioned above, Eq. 1.19 is derived from elastic theorem, so Eqs. 1.20 and 1.23 are suitable only for elastic range too. Nevertheless, for a plastic material, it is impossible for the stress near crack tip to increase infinitely when the stress reaches yield limit σ_s. When the material in a small area near crack tip is in a plastic state (shown in Fig. 1.13) there will inevitably be a stress distribution. As a result, it is essential to correct Eqs. 1.20 and 1.23 by considering the effect of plastic zone, i.e., assuming that a plastic zone with a radius of r'_p exists near crack tip in non-plastic zone, then the variable of $2a$ in Eqs. 1.20 and 1.23 should be substituted with the variable of $2(a + r'_p)$, termed as the effective crack length and expressed through $2a_1$ as:

$$2a_1 = 2\left(a + r'_p\right) \quad (1.27)$$

Table 1.1 Correction factors of stress intensity factor of a plate with a finite width

Types	Correction factor α
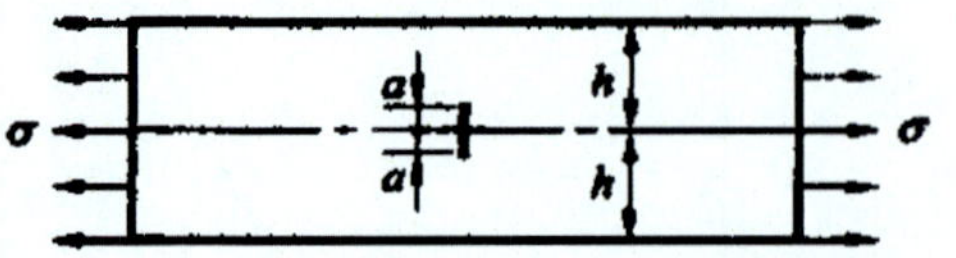	$\alpha = \frac{1}{\sqrt{\pi}}\left[1.77 + 0.277\left(\frac{a}{h}\right) - 0.51\left(\frac{a}{h}\right)^2 + 2.7\left(\frac{a}{h}\right)^3\right]$ $\alpha = 1.00$ (in the case of $a \ll h$)
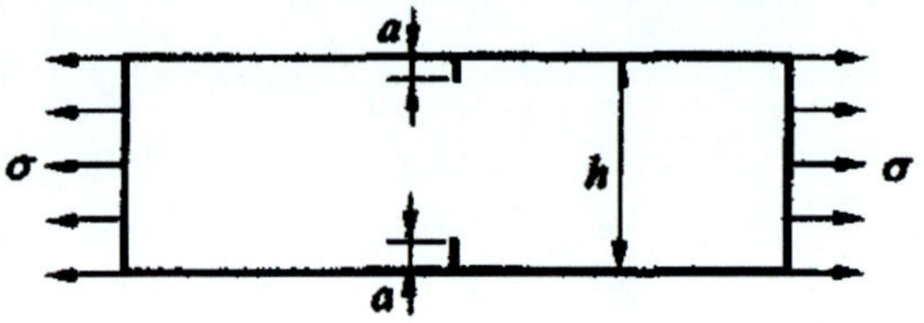	$\alpha = \frac{1}{\sqrt{\pi}}\left[1.99 - 0.41\left(\frac{a}{h}\right) + 18.7\left(\frac{a}{h}\right)^2 - 38.48\left(\frac{a}{h}\right)^3 + 53.85\left(\frac{a}{h}\right)^4\right]$ $\alpha = 1.12$ (in the case of $a \ll h$)
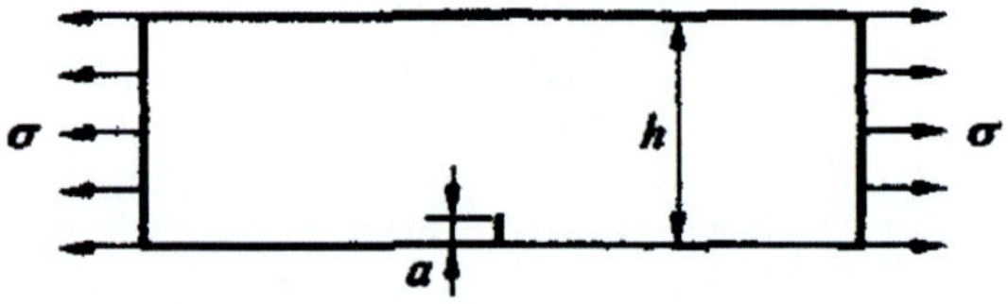	$\alpha = \frac{1}{\sqrt{\pi}}\left[1.99 + 0.38\left(\frac{2a}{h}\right) - 2.12\left(\frac{2a}{h}\right)^2 + 3.42\left(\frac{2a}{h}\right)^3\right]$ $\alpha = 1.12$ (in the case of $a \ll h$)
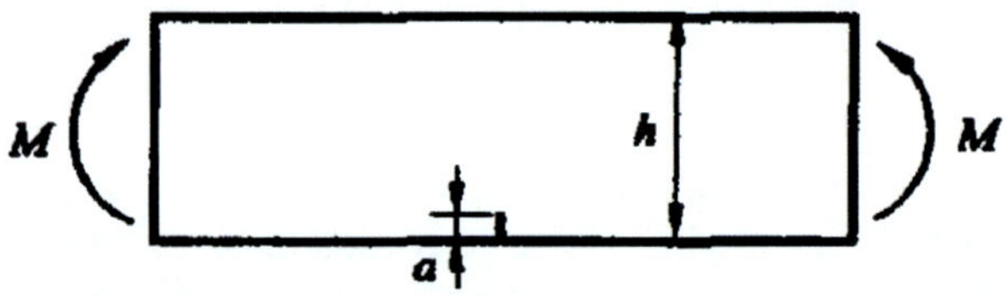	$\alpha = \frac{1}{\sqrt{\pi}}\left[1.99 - 2.47\left(\frac{a}{h}\right) + 12.97\left(\frac{a}{h}\right)^2 - 23.17\left(\frac{a}{h}\right)^3 + 24.80\left(\frac{a}{h}\right)^4\right]$ $\alpha = 1.12$ (in the case of $a \ll h$)

$\sigma = \frac{6M}{bh^2}$, here b is the thickness of plate

where $r'_p = r_p \cdot F_s$. r_p is the width of plastic zone along the extension line of crack. F_s is the correction factor considered the actual shape of plastic zone.

Based on plastic zone theorem, for plane stress state, one has:

$$r_p = \frac{\bar{K}_1^2}{2\pi\sigma_s^2} \tag{1.28}$$

Substituting Eq. 1.26 into Eq. 1.28, it can be show that:

$$r_p = \frac{a}{2}\left(\frac{\sigma}{\sigma_s}\right)^2 \alpha^2 \tag{1.29}$$

Letting $Z = \sigma/\sigma_s$, then Eqn. (1.29) becomes:

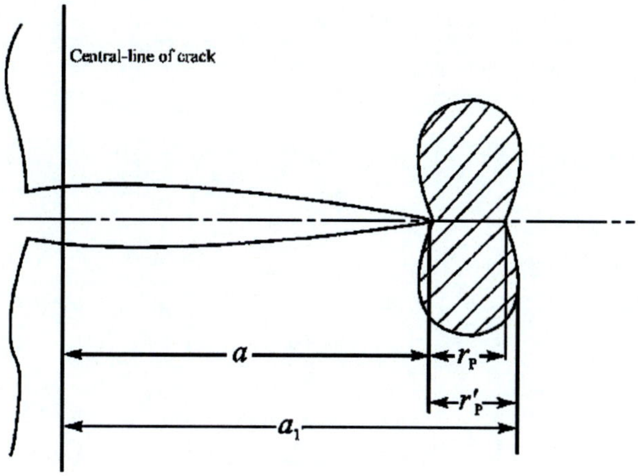

Fig. 1.13 Shape of plastic zone

$$r_p = \frac{a}{2}Z^2\alpha^2 \tag{1.30}$$

Substituting Eq. 1.30 into Eq. 1.27, it is possible to have:

$$a_1 = a + r'_p = a + r_p F_s = a\left(1 + \frac{1}{2}Z^2\alpha^2 F_s\right) \tag{1.31}$$

Again letting $k = 1 + \frac{1}{2}Z^2\alpha^2 F_s$, then Eq. 1.31 can be written as:

$$a_1 = ka \tag{1.32}$$

From plastic zone theorem, it is also well known that:

$$F_s = 1 + \frac{4}{\pi}\left(\frac{Z^2}{1 - Z^2}\right) \tag{1.33}$$

Taking the correction of plastic zone into account, then the stress intensity factor for a thin plate (plane strain state) with finite width and central crack becomes:

$$\bar{K}_I = \sqrt{\pi a_1} \cdot \sigma \cdot \alpha \tag{1.34}$$

As the plastic zone and plastic width r_p near crack tip under plane strain state are much lower than those under plane stress state, generally, plastic width r_p under plane strain state can be determined by using following empirical formula as:

$$r_p = \frac{K_I^2}{6\pi\sigma_s^2} \tag{1.35}$$

Thus, for plane strain state, by aid of Eqs. 1.32–1.34, one has:

$$K_I = \sqrt{\pi a_1} \cdot \sigma \cdot \alpha \tag{1.36}$$

$$a_1 = ka$$

$$k = 1 + \frac{1}{6} Z^2 \alpha^2 F_s$$

$$F_s = 1 + \frac{4}{\pi}\left(\frac{Z^2}{1 - Z^2}\right)$$

1.6 Crack Growth Rate and Generalized Fracture S–N Surface

In damage tolerance design, the damage would be temporarily tolerated but there needs to be enough residual strength (i.e., fail-safe load) in the component for it to keep for resisting against fracture failure until eventual repair. From the different demands of structural components for different services in aircraft and ships, in general, fail-safe load amounts to about 60 ~ 80% of ultimate load. As shown in Fig. 1.14, an un-notched component (i.e., $a = 0$) may carry the ultimate load whereas a notched part with an initial surface crack of a_0, subjected to a static critical stress of σ_c (or the stress intensity factor near crack tip reaches the critical value of K_{IC} (or K_C)), would have failed (shown in Fig. 1.15). With the decreasing of static stress to σ_0, the component would not have failed. However, under a

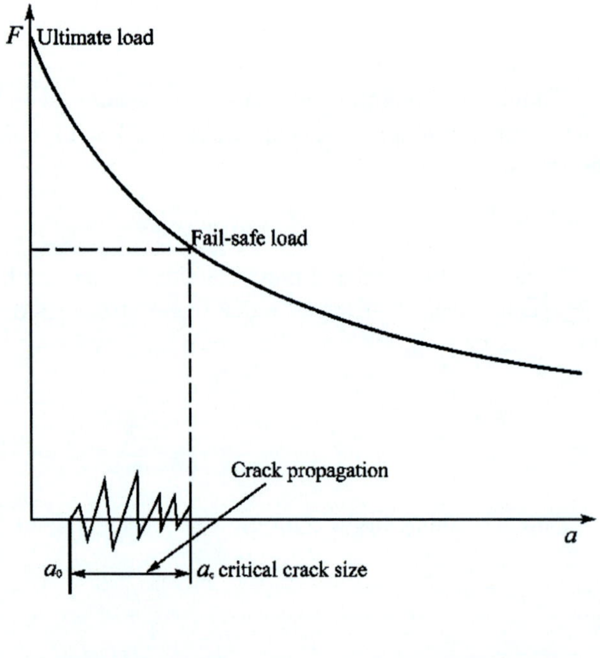

Fig. 1.14 *P–a* curve of crack growth

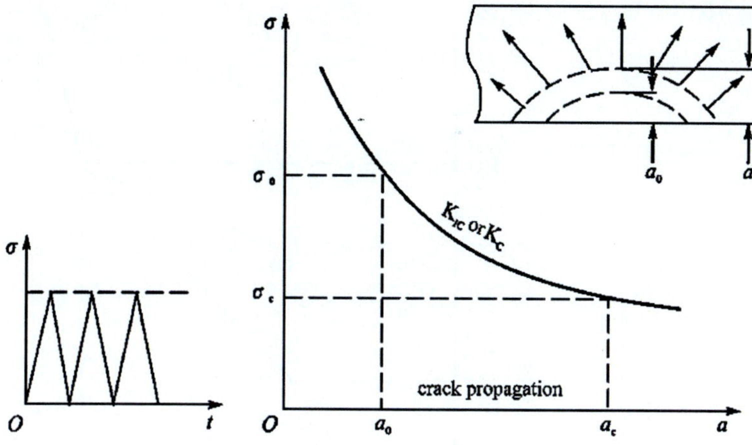

Fig. 1.15 Critical crack size and subcritical crack propagation

cyclic stress regime with the same value as the static stress, i.e. σ_0 (shown in the left of Fig. 1.15), after the component has an initial crack of a_0, the crack would grow until it reaches a critical length of a_c and brittle facture would then occur. The crack size a_c relevant to the fail-safe load is called as the critical crack size. Under cyclic stress, the process of crack propagation from an initial value of a_0 to a critical value of a_c is termed as the subcritical growth of fatigue crack and the time of crack propagation from a detectable size of a_0 to a critical size of a_c is known as the crack propagation life. In order to assure the safe operation of a component, it is desirable for the crack propagation life of a component to be greater than the inspection period. Then an important problem arising from damage tolerance design is how the crack propagation life of component should be determined under a fatigue load spectrum.

Preconditions to obtaining fatigue crack propagation life include knowing the crack growth rate, detectable crack size a_0 and the final or critical crack size a_c. In general, the size a_0 is dependent on the non-destructive examination (NDE) method for structures in service, a reasonable limit for a good NDE method is about 1.0 mm. The critical crack size a_c could be determined by using the *P–a* curve (shown in Fig. 1.14), which is usually obtained from static strength experiments from a set of specimens with different crack sizes *a*. However, this is not an economic and time-efficient method. Therefore, the fracture mechanics point of view is employed for the tests. As mentioned in the above section, the fracture criterion of notched component is:

$$\bar{K}_{\mathrm{I}} = K_{\mathrm{C}} \tag{1.37}$$

and:

$$\bar{K}_{\mathrm{I}} = \sqrt{\pi a_1} \cdot \sigma \cdot \alpha = \sqrt{\pi k a} \cdot \sigma \cdot \alpha = \left(\sqrt{\pi k} \cdot \alpha\right) \sigma \cdot \sqrt{a}$$

Thus:

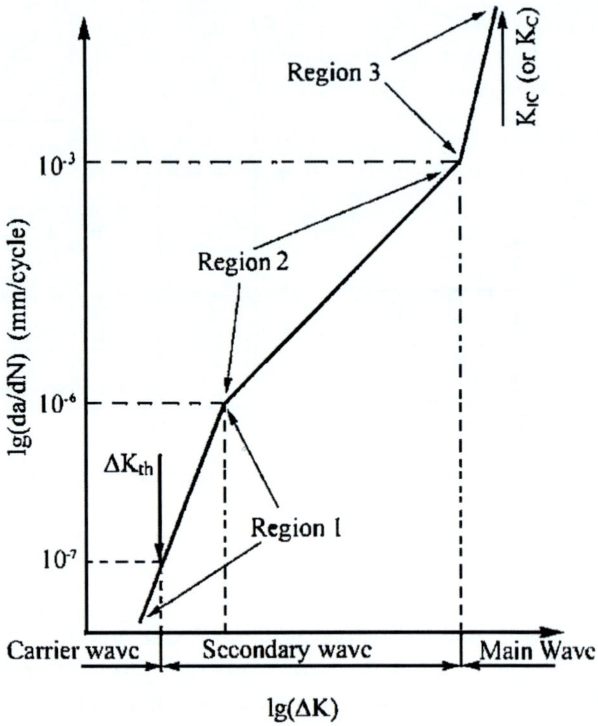

Fig. 1.16 *da/dN*–Δ*K* curve

$$\left(\sqrt{\pi k}\cdot\alpha\right)\sigma\cdot\sqrt{a} = K_C$$

and then:

$$\sigma = \frac{K_C}{\sqrt{\pi k}\cdot\alpha}\cdot\frac{1}{\sqrt{a}} \tag{1.38}$$

Equation 1.38 depicts a σ–a curve, from which a P–a curve can be obtained through transformation. It can be shown that in the case $a = 0$, then $P \to \infty$; evidently, this is inconsistent with the practical situation. Thus, as shown in Fig. 1.14, in the case of $a = 0$, then $P = P_b$ (ultimate load) and the P–a curve from $a = 0$ can be regarded as the reference base for fracture tests.

Fatigue cracks generally appear on the surface of material or at large inclusions, resulting from high stresses, surface roughness, fretting, corrosion, etc. Fatigue crack growth on a macroscopic level usually occurs perpendicular to the main or principal stress and is dependent on the material, the material thickness and the orientation of the crack relative to principal material directions. Furthermore, the crack growth depends on the cyclic stress amplitude, the mean stress and the environment. The crack growth rate, denoted *da*/*dN*, has become an important "material property" to characterize fatigue crack propagation under constant amplitude loading. Normally, three regions of crack growth rate are identified as shown in Fig. 1.16. Region 1 is usually referred to as the near threshold region owing to the threshold stress intensity range, ΔK_{th}, below which fatigue crack growth will not occur. This is believed to be true for many materials but for some material-environment combinations the slope of the *da*/*dN* versus ΔK relationship has been found to be finite even for growth rates as low as 10^{-7} mm/cycle. Region 2 is usually referred to as the stable or linear crack growth rate region, since the Paris relation usually fits the data in this region very well. Finally, region 3 is often referred to as the unstable crack growth region, since crack growth rate increases very rapidly as the maximum stress intensity factor approaches the fracture toughness K_C.

It is well known that the crack growth rate is presented as a function of the stress intensity factor range ΔK for different stress ratios R, material thicknesses, and different environments. Various deterministic fatigue crack growth rate functions have been proposed in the literature. The functions can be represented by a general form [10, 11]:

$$\frac{da(t)}{dt} = F(\Delta K, K_{\max}, R, S, a) \tag{1.39}$$

where a is the crack length, $a(t)$ is the crack length dependent on time t, ΔK is the stress intensity factor range, $K_{\max}$ is the spectrum peak stress intensity factor, S is the fatigue strength, or stress amplitude, or peak stress level in the loading spectrum, da/dt is crack growth per cycle and $F(\Delta K, K_{\max}, R, S, a)$ is a non-negative function. Some crack growth rate functions, such as Paris-Erdogan model [12], Trantina-Johnson model [13], Walker model [14], Forman model [15], and generalized Forman model [16], are commonly used.

(1) Paris-Erdogan model [12]:

$$\frac{da}{dN} = C(\Delta K)^n \tag{1.40}$$

(2) Trantina-Johnson model [13]:

$$\frac{da}{dN} = C(\Delta K - \Delta K_{th})^n \tag{1.41}$$

(3) Walker model [14]:

$$\frac{da}{dN} = C(\Delta K)^n (1 - R)^m \tag{1.42}$$

(4) Forman model [15]:

$$\frac{da}{dN} = C\frac{(\Delta K)^n}{(1 - R)K_C - \Delta K} \tag{1.43}$$

(5) Generalized Forman model [16]:

$$\frac{da}{dN} = C\left[\left(\frac{1-f_0}{1-R}\right)\Delta K\right]^{m_1} \frac{\left(1-\frac{\Delta K_{th}}{\Delta K}\right)^{m_2}}{\left[1-\frac{\Delta K}{(1-R)K_C}\right]^{m_3}} \tag{1.44}$$

where C, n, m, m_1, m_2 and m_3 are the material constants, ΔK_{th} is the crack threshold, K_C is the plane stress fracture toughness of material dependent on the thickness of structures, but for a general thickness about 1.0–2.5 mm of thin plate, K_C is approximated to be dependent only on the material and f_0 is the fatigue crack opening function, which can be determined as [16]

$$f_0 = \frac{K_{open}}{K_{max}} = \begin{cases} \max(R, A_0 + A_1R + A_2R^2 + A_3R^3) & R \geq 0 \\ A_0 + A_1R & -2 \leq R < 0 \end{cases} \tag{1.45}$$

with:

$$A_0 = \left(0.825 - 0.34\alpha_0 + 0.05\alpha_0^2\right)\left[\cos\left(\frac{\pi S}{2\sigma_0}\right)\right]^{\frac{1}{\alpha_0}}$$

$$A_1 = (0.415 - 0.071\alpha_0)\frac{S}{\sigma_0}$$

$$A_2 = 1 - A_0 - A_1 - A_3$$

$$A_3 = 2A_0 + A_1 - 1$$

where α_0 is the plane stress/strain constraint factor, and S/σ_0 is the ratio of maximum stress to the flow stress.

Taking the transformation and integration of Eq. 1.39 yields:

$$N^* = \int_{a_0}^{a_c} \frac{1}{F(\Delta K, K_{max}, R, S, a)} da \tag{1.46}$$

where N^* represents the fatigue crack growth life and the right superscript character of "*" is used for distinguishing the fatigue crack growth life from the fatigue crack formation life. Equation 1.46 describes the relationship between the stress S and fatigue crack growth life N^* and is known as the crack growth S–N curve.

According to elastic fracture mechanics, in general, the stress intensity factor K in Eqs. 1.39–1.44 is the product of both functions concerning stress S and crack size a:

$$K = X(S) \cdot Y(a) \tag{1.47}$$

where $Y(a) = \alpha(a)\sqrt{\pi a}$ and $\alpha(a)$ is the geometry correction function of fatigue crack.

From Eq. 1.47, the stress intensity factor mean and range become respectively:

$$K_m = X(S_m) \cdot Y(a) \tag{1.48}$$

$$\Delta K = K_{max} - K_{min} = [X(S_{max}) - X(S_{min})] \cdot Y(a) = \Delta X \cdot Y(a) \tag{1.49}$$

In the case of considering the correction of plastic zone near crack tip, then:

$$X(S) = \frac{S}{\sqrt{1 - \pi\beta(S/\sigma_s)^2}}$$

where σ_s is the yield limit. β is the material constant, under plane stress state, $\beta = 1/(2\pi)$ and under plane strain state, $\beta = (1-2v)^2/(2\pi)$, here v is the Poisson ratio.

Since a fatigue stress cycle is defined by two stress components of amplitude S_a and mean S_m, and $S_{max} = S_m + S_a$, $S_{min} = S_m - S_a$, then:

$$\Delta X = X(S_{max}) - X(S_{min}) = \frac{S_m + S_a}{\sqrt{1 - \pi\alpha[(S_m + S_a)/\sigma_s]^2}} - \frac{S_m - S_a}{\sqrt{1 - \pi\alpha[(S_m - S_a)/\sigma_s]^2}} \tag{1.50}$$

Substituting Eq. 1.50 into Eq. 1.49, one has:

$$\Delta K = \left\{ \frac{S_m + S_a}{\sqrt{1 - \pi\alpha[(S_m + S_a)/\sigma_s]^2}} - \frac{S_m - S_a}{\sqrt{1 - \pi\alpha[(S_m - S_a)/\sigma_s]^2}} \right\} Y(a) \tag{1.51}$$

In the case of without considering the correction of plastic zone near crack tip, Eqs. 1.48 and 1.51 degenerate into:

$$K_m = S_m \cdot Y(a) \tag{1.52}$$

$$\Delta K = 2S_a Y(a) \tag{1.53}$$

Substituting Eqs. 1.52 and 1.53 into Eq. 1.44, and from Eq. 1.46, it is possible to have:

$$N^* = \frac{(2S_a)^{m_2}}{C \cdot [(1-f)(S_a + S_m)]^{m_1}} \int_{a_0}^{a_{cr}} \frac{[Y(a)]^{(m_1 - m_2)} \cdot [2S_a Y(a) - \Delta K_{th}]^{m_2}}{\left[1 - \frac{S_a + S_m}{K_C} Y(a)\right]^{m_3}} da \tag{1.54}$$

Equation 1.54 reveals the relationship between stress amplitude S_a, mean stress S_m and fatigue crack growth life N^*, displaying an S_a–S_m–N^* surface in three-dimensional coordinate system. The analogy of Eq. 1.54 can be deployed to obtain the generalized fracture S–N surfaces pertinent to Eqs. 1.40–1.43.

1.7 Total Life Prediction

As mentioned in Sect. 1.5, the fatigue process includes two-stages of crack initiation and crack propagation and total fatigue life of a structural component is then equal to the sum of the time to crack initiation (i.e., to visually detectable crack size) and the time to crack propagation until its critical size. From cumulative damage Miner's rule, fatigue crack formation and propagation lives can be predicted respectively by using generalized fatigue and fracture S–N surfaces (1.16) and (1.54).

In the case that a block of complex actual load spectrum includes k stress levels of $s_1, s_2, \ldots, s_k$, pertaining to the stress cycle numbers of $n_1, n_2, \ldots, n_k$ respectively, with $N_1, N_2, \ldots, N_k$ representing the loading cycles to failure under the independent loading of the above stress levels respectively, then, based on linear cumulative damage Miner's theory, fatigue damage can be expressed as the ratio of cycle, or $n_1/N_2, n_2/N_2, \ldots, n_k/N_k$. If T denotes the number of blocks of actual load spectrum, each stress level in this block of actual load spectrum respectively cause the damages of components as:

$$T\frac{n_1}{N_1}, T\frac{n_2}{N_2}, \ldots, T\frac{n_L}{N_L} \tag{1.55}$$

When the sum of damage is accumulated to be 1 (or 100%), i.e.,

$$T \cdot \sum_{i=1}^{k} \frac{n_i}{N_i} = 1 \tag{1.56}$$

Then a fatigue crack is deemed to have initiated. Under the cyclic loading of sequential and stochastic spectra, Eq. 1.56 can be written into an integral form:

$$T \cdot \int_{0}^{(s)_{\max}} \frac{n_T f(s)}{N(s)} ds = 1 \tag{1.57}$$

where n_T represents the total frequency of cyclic stress in a block of load spectrum, $f(s)$ is the probability density function of cyclic stress of s and $N(s)$ represents the cycle number to fatigue crack initiation under the independent loading of cyclic stress of s.

It is well-known that the cyclic stress level in a block of load spectrum is dominated by the two-parameter nominal stress (s_a, s_m), Eqs. 1.56 and 1.57 can be therefore extended to the two-parameter stress forms as:

$$T \cdot \sum_{j=1}^{k} \sum_{i=1}^{h} \frac{n_{ij}(s_a, s_m)}{N_{ij}(s_a, s_m)} = 1 \tag{1.58}$$

$$T \cdot \int_{(s_m)_{min}}^{(s_m)_{max}} \int_{(s_a)_{min}}^{(s_a)_{max}} \frac{n_T f(s_a, s_m)}{N(s_a, s_m)} \mathrm{d}s_a \mathrm{d}s_m = 1 \tag{1.59}$$

where $n_{ij}(s_a, s_m)$ is the number of cyclic stress of (s_a, s_m) in a block of load spectrum. $f(s_a, s_m)$ represents the probability density function of cyclic stress of (s_a, s_m) and $N_{ij}(s_a, s_m)$ is the cycle number to fatigue crack initiation under the independent loading of cyclic stress of (s_a, s_m).

Actually, Miner's rule expressed in Eqs. 1.56–1.59 contains model uncertainty. In order to predict fatigue crack formation life exactly, Eqs. 1.58 and 1.59 can be written as:

$$T \cdot \sum_{j=1}^{k} \sum_{i=1}^{h} \frac{n_{ij}(s_a, s_m)}{N_{ij}(s_a, s_m)} = a \tag{1.60}$$

$$T \cdot \int_{(s_m)_{min}}^{(s_m)_{max}} \int_{(s_a)_{min}}^{(s_a)_{max}} \frac{n_T f(s_a, s_m)}{N(s_a, s_m)} ds_a ds_m = a \tag{1.61}$$

where a is determined from the experiments of like structures. Equations 1.58–1.61 are applied only for the load spectrum without large overload and have a good correlation to the experiments under steady cyclic loading without significant change.

Using the analogy of Eqs. 1.58–1.61, it can be obtained Miner's rules for predicting fatigue crack propagation life as:

$$T^* \cdot \sum_{j=1}^{k} \sum_{i=1}^{h} \frac{n_{ij}(s_a, s_m)}{N^*_{ij}(s_a, s_m)} = 1 \tag{1.62}$$

$$T^* \cdot \int_{(s_m)_{min}}^{(s_m)_{max}} \int_{(s_a)_{min}}^{(s_a)_{max}} \frac{n_T f(s_a, s_m)}{N^*(s_a, s_m)} ds_a ds_m = 1 \tag{1.63}$$

$$T^* \cdot \sum_{j=1}^{k} \sum_{i=1}^{h} \frac{n_{ij}(s_a, s_m)}{N^*_{ij}(s_a, s_m)} = a \tag{1.64}$$

$$T^* \cdot \int_{(s_m)_{min}}^{(s_m)_{max}} \int_{(s_a)_{min}}^{(s_a)_{max}} \frac{n_T f(s_a, s_m)}{N^*(s_a, s_m)} ds_a ds_m = a \tag{1.65}$$

where $N^*(s_a, s_m)$ can be determined from generalized fracture S–N surfaces (1.54).

References

1. Goodman J (1899) Mechanics applied to engineering (1st ed.). London, Longmans, Green and Co

2. Gerber WZ (1874) Bestimmung der zulässigen Spannungen in Eisen-Constructionen. [Calculation of the allowable stresses in iron structures]. Z Bayer Archit Ing Ver 6(6):101–110
3. Soderberg CR (1930) Factor of safety and working stress. Transaction of American Society of Mechanical Engineering, Part APM-52-2, 13–28
4. Zhao S, Wang Z (1992) Fatigue design. Mechanical Industry Press, Beijing, pp 48–51 (in Chinese)
5. Bagci C (1981) Fatigue design of machine elements using the 'Bagci line' defining the fatigue failure surface line (mean stress diagram). Mech Mach Theory 16(4):339–359
6. Kujawski D, Ellyin F (1995) A unified approach to mean stress effect on fatigue threshold conditions. Int J Fatigue 17:101–106
7. Xiong JJ, Shenoi RA, Zhang Y (2008) Effect of the mean strength on the endurance limit or threshold value of the crack growth curve and two-dimensional joint probability distribution. J Strain Anal Eng Des 43(4):243–257
8. Wheeler OE (1972) Spectrum loading and crack growth. J Basic Eng 94:181–186
9. Willenborg J, Engle RM, Wood HA (1971) A crack growth retardation model using an effective stress concept. Report No. AFFDL-TR71-1. Air Force Flight Dynamic Laboratory. Wright-Patterson Air Force Base, USA
10. Miller MS, Gallagher GP (1981) An analysis of several fatigue crack growth rate descriptions. Measurement and Data Analysis, ASTM STP 738, 205–251
11. Hoeppner DW, Krupp WE (1974) Prediction of component life by application of fatigue crack growth knowledge. Eng Fract Mech 6:47–70
12. Paris PC, Erdogan F (1963) A critical analysis of crack propagation laws. J Basic Eng Transaction ASME (Series D) 85:528–534
13. Trantina GG, Johnson CA (1983) Probabilistic defect size analysis using fatigue and cyclic crack growth rate data. Probabilistic Fracture Mechanics and Fatigue Methods, ASTM STP 798, 67–78
14. Walker EK (1970) The effect of stress ratio during crack propagation and fatigue for 2024-T3 and 7075-T6 aluminum. Effects of Environment and Complex Load History on Fatigue Life, ASTM STP 462:1–14
15. Forman RG et al (1967) Numerical analysis of crack propagation in cyclic loaded structures. J Basic Eng Transaction ASME (Series D) 89:459–465
16. Newman JC Jr (1984) A crack opening equation for fatigue crack growth. Int J Fract 24:131–135

Chapter 2
Reliability and Confidence Levels of Fatigue Life

2.1 Introduction

As is well known, fatigue lives of nominally identical specimens subjected to the same nominal cyclic stress display scatter as shown schematically in Fig. 2.1. This phenomenon reflects the stochastic nature of a fatigue damage process. Previous researches reveal that the uncertainty modeled by the stochastic variables can be divided in the following groups [1, 2]: (1) physical uncertainty, or inherent uncertainty, that is related to the natural randomness of a quantity, (2) measurement uncertainty, i.e., the uncertainty caused by imperfect measurements, (3) statistical uncertainty, which is due to limited sample sizes of observed quantities, (4) model uncertainty, one related to imperfect knowledge or uncertain idealizations of the mathematical models used or uncertainty related to the choice of probability distribution types for the stochastic variables. Based on this, many stochastic mathematical expressions for fatigue damage process have been developed. Due to the variations between individual specimens, fatigue data can be described by random variables to study the variability of fatigue damage and life and to analyze their average trends. With improvement in crack-size measurements, fatigue crack growth data can be depicted by random fields/stochastic processes in a random time–space and state-space to indicate local variations within a single specimen and to analyze the statistical nature of fatigue crack growth data. This has been done by a stationary lognormal process-based randomized approach of deterministic crack growth equation in power law and polynomial forms. In order to understand the stochastic nature of fatigue damage characterization and statistically meaningful data sets, it is desirable to have a technique that accounts for small sample numbers to determine structural fatigue life and performance, which is the focus of this chapter.

J. J. Xiong and R. A. Shenoi, *Fatigue and Fracture Reliability Engineering*,
Springer Series in Reliability Engineering, DOI: 10.1007/978-0-85729-218-6_2,

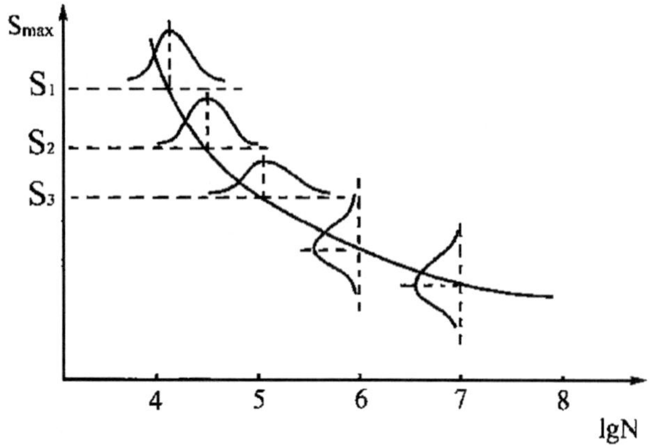

Fig. 2.1 S–N curve

2.2 Basic Concepts in Fatigue Statistics

It is well known that fatigue is a two-stage process, those of crack initiation and crack propagation; total fatigue life of a structural component is then equal to the sum of crack initiation and crack propagation lives. In general, fatigue life is governed by loading in three interval ranges as follows

(1) Low life region, i.e., under large strain cycles, fatigue life of a specimen is less than about 10^4 cycles;
(2) Medium life region, with fatigue life of a specimen amounts from 10^4 to 10^6 cycles;
(3) Long life region, where under low stress cycles, fatigue life of specimen is greater than about 10^6 cycles.

In general, fatigue lives in the long life region show a greater dispersion than those in the low life region. The factors influencing the dispersion of fatigue experiments, that are also termed as occasional factors, include: (1) measurement equipment uncertainty, (2) inhomogeneity of experimental material, i.e., the test specimens were cut along different orientations of the primary material, (3) inconsistency of specimen dimension and configuration, (4) inconsistency of specimen processing procedures, (5) variability of specimen during a heat-treating process, e.g., different positions of specimens in heat-treating furnace, (6) occasional changes in the experimental environment.

Well planned experiments are necessary to determine how fatigue life data are influenced by these occasional factors. The test variables, e.g., fatigue life, fatigue load, fatigue limit and strength limit, etc., dependent on these kinds of occasional factors, are termed as the random variables.

The population represents all subjects investigated, whereas the individual indicates a basic unit in a population. The behaviour of the population is dependent on the behaviour of the many individuals. Then it is essential to understand the behaviour of each individual to obtain the characteristics of the population. In this there are two primary problems whereby: (1) a population is generally composed of a so large number on individuals, even infinite, that it is impossible to investigate all individuals. (2) for a few full-scale parts in industrial production, fatigue tests for determining fatigue lives of individuals are destructive since the tested parts cannot then be practical use, i.e. it is infeasible to perform destructive tests of whole parts.

In general, some individuals are randomly sampled from the population to be tested for inferring the nature of population. These sampled individuals are called as the sample and the number of individuals in a sample is termed as the sample size. The components and parts for fatigue tests, or the small standard coupons for determining fatigue behaviour of material are generally known as the specimens. A determined value of fatigue life of a specimen refers to an individual and an experimental dataset of a set of specimens refers to a sample. For example, when sample size equals five, this means that the sample includes five observed values.

The eigenvalues of the observed data representing statistical nature of a sample may be classified into two categories as: (1) the central position of data, e.g., mean and median, (2) the dispersion of data, including standard deviation, variance and coefficient of variation, etc. If a sample with a sample size of n is randomly sampled from a population to obtain n observed values of $x_1, x_2, \ldots, x_n$, then the mean of n observed values is the sample mean and is denoted as

$$\bar{x} = \frac{1}{n}\sum_{i=1}^{n} x_i \tag{2.1}$$

The sample mean represents the central position of data.

Besides the arithmetic mean $\bar{x}$ of sample, the geometric mean G of sample is also usually used in fatigue reliability analysis. In the case of n observed values of $x_1, x_2, \ldots, x_n$, then the geometric mean G of sample is

$$G = \left(\prod_{i=1}^{n} x_i\right)^{1/n} \tag{2.2}$$

where $\prod$ is the continued multiplication notation and $\prod_{i=1}^{n} x_i$ represents the continued multiplication of n observed values of $x_1, x_2, \ldots, x_n$. The logarithmic form of Eq. 2.2 becomes

$$\log G = \frac{1}{n}\sum_{i=1}^{n} \log x_i \tag{2.3}$$

From Eq. 2.3, it is clear that the logarithm of geometric mean G equals to the arithmetic mean of logarithm of each observed value. In general usage, the mean implies the arithmetic one.

The median is also a characteristic value to depict the central position of data. Taking a set of data into sequential arrangement, then the mid value is called as the sample median of the dataset and denoted as M_e. In the case of odd number of observed data the sample median is the mid value, while in the case of even number of observed data the sample median is the mean of two mid values.

As a way of measuring dispersion, the sample variance s^2 is defined as

$$s^2 = \frac{\sum_{i=1}^{n} (x_i - \bar{x})^2}{n - 1} \tag{2.4}$$

Or alternatively,

$$s^2 = \frac{\sum_{i=1}^{n} x_i^2 - \frac{1}{n}\left(\sum_{i=1}^{n} x_i\right)^2}{n - 1} \tag{2.5}$$

where n is the number of observed values; $(n - 1)$ is the freedom degree of variance. $\sum_{i=1}^{n} x_i^2$ is the sum of squares of observed value and $\left(\sum_{i=1}^{n} x_i\right)^2$ is the squares of sum of observed value.

The standard deviation is another characteristic value to describe the dispersion of observed data. The square root s of sample variance s^2 is termed as the sample standard deviation, namely

$$s = \sqrt{\frac{\sum_{i=1}^{n} (x_i - \bar{x})^2}{n - 1}} \tag{2.6}$$

or

$$s = \sqrt{\frac{\sum_{i=1}^{n} x_i^2 - \frac{1}{n}\left(\sum_{i=1}^{n} x_i\right)^2}{n - 1}} \tag{2.7}$$

The formulations of variance and standard deviation can also be written as

$$s^2 = \frac{\sum_{i=1}^{n} x_i^2 - n\bar{x}^2}{n - 1} \tag{2.8}$$

$$s = \sqrt{\frac{\sum_{i=1}^{n} x_i^2 - n\bar{x}^2}{n - 1}} \tag{2.9}$$

Some characteristic features are as follows:

(1) The standard deviation is an important index to indicate the dispersion of data, i.e., a greater standard deviation means a larger dispersion of observed data.
(2) The standard deviation is positive with the same unit of observed value.
(3) The standard deviation for a set of observed values, randomly sampled from the population, is termed as the sample standard deviation, which is different from the population standard deviation mentioned below.

The standard deviation is calculated through the deviations of observed values from the mean. It depends only on the absolute deviation of each observed value and is independent on the absolute value of each observed data. In order to consider the influence of the observed value on the standard deviation, dividing the standard deviation by the mean yields the characteristic value, namely, the coefficient of variation or coefficient of dispersion C_v as

$$C_v = \frac{s}{\bar{x}} \times 100\%$$

The coefficient of variation is an important index to indicate the relative dispersion of a dataset; it is a dimensionless unit and is generally used for comparing dispersions between two sets of observed values with possibly different features and units.

As mentioned above, fatigue life, fatigue load, fatigue limit, strength limit, etc., are random variables, whose expected value ξ, is defined as

$$E(\xi) = \int_{-\infty}^{\infty} xf(x)dx \tag{2.10}$$

$E(\xi)$ represents the central position of the random variable distribution. If the variance of random variable ξ is denoted as $\mathrm{Var}(\xi)$, then this is

$$\mathrm{Var}(\xi) = \int_{-\infty}^{\infty} [\mathrm{x} - \mathrm{E}(\xi)]^2 \mathrm{f}(\mathrm{x})\mathrm{dx} \tag{2.11}$$

The square root $\sqrt{\mathrm{Var}(\xi)}$ is termed as the standard deviation of random variable ξ. The expression of variance $\mathrm{Var}(\xi)$ can be simplified into

$$\mathrm{Var}(\xi) = \mathrm{E}(\xi^2) - 2\mathrm{E}(\xi)\mathrm{E}(\xi) + [\mathrm{E}(\xi)]^2$$

Thus,

$$\mathrm{Var}(\xi) = \mathrm{E}(\xi^2) - [\mathrm{E}(\xi)]^2 \tag{2.12}$$

No matter which distributions two random variables ξ and η follow, and whether ξ and η are mutually independent or not, the mathematical expectation of $\xi + \eta$ is equivalent to the sum of the mathematical expectations of ξ and η. If $E(\xi)$ and $E(\eta)$ are known, then it can be shown that

$$E(\xi + \eta) = E(\xi) + E(\eta) \tag{2.13}$$

Similarly, it is possible to have the mathematical expression for the difference between two random variables ξ and η as

$$E(\xi - \eta) = E(\xi) - E(\eta) \tag{2.14}$$

Using Eq. 2.13 as an analogy, one can write the sum of the central positions of n random variables $\xi_1, \xi_2, \ldots, \xi_n$ as

$$E(\xi_1 + \xi_2 + \cdots + \xi_n) = E(\xi_1) + E(\xi_2) + \cdots + E(\xi_n) \tag{2.15}$$

The variance of the sum of random variables ξ and η may be obtained as

$$\mathrm{Var}(\xi+\eta)=\mathrm{Var}(\xi)+\mathrm{Var}(\eta)+2\mathrm{Cov}(\xi,\eta) \tag{2.16}$$

In the case where two random variables ξ and η are mutually independent, it can be proved that the covariance of random variables ξ and η equals to zero, or $Cov(\xi,\eta)=0$. Then Eq. 2.16 becomes

$$\mathrm{Var}(\xi+\eta)=\mathrm{Var}(\xi)+\mathrm{Var}(\eta) \tag{2.17}$$

Again, using the analogy of Eqs. 2.16 and 2.17, it is possible to have

$$\mathrm{Var}(\xi-\eta)=\mathrm{Var}(\xi)+\mathrm{Var}(\eta)-2\mathrm{Cov}(\xi,\eta) \tag{2.18}$$

$$\mathrm{Var}(\xi-\eta)=\mathrm{Var}(\xi)+\mathrm{Var}(\eta) \tag{2.19}$$

From Eqs. 2.17 and 2.19, it can be concluded that no matter which distribution two random variables ξ and η follow, the variance of random variable $\xi+\eta$ is equivalent to that of $\xi-\eta$ and equals to $\mathrm{Var}(\xi)+\mathrm{Var}(\eta)$. However, in the case of two dependent random variables, it is necessary to know probability density function (PDF) $p(\xi, \eta)$ of two-dimensional random variables (ξ, η) to obtain the covariance $Cov(\xi, \eta)$ and to determine the variance of the sum (or difference) between two-dimensional random variables.

Again, using the analogy of Eq. 2.17, it is possible to obtain the variance of the sum of n random variables ξ_1, ξ_2, ..., ξ_n, where if n random variables are mutually independent, then

$$\mathrm{Var}(\xi_1+\xi_2+\cdots+\xi_\mathrm{n})=\mathrm{Var}(\xi_1)+\mathrm{Var}(\xi_2)+\cdots+\mathrm{Var}(\xi_\mathrm{n}) \tag{2.20}$$

Since the sample mean $\bar{\xi}$ is a random variable function as

$$\bar{\xi}=\frac{1}{n}(\xi_1+\xi_2+\cdots+\xi_n)=\frac{1}{n}\sum_{i=1}^{n}\xi_i$$

from Eq. 2.15, it can be deduced that

$$E(\bar{\xi})=\frac{1}{n}[E(\xi_1)+E(\xi_2)+\cdots+E(\xi_n)]$$

As all individuals (observed values) in a sample come from a same population, random variables ξ_1, ξ_2, ...ξ_n have the same PDF. Letting the mean of their same population be μ, i.e.,

$$E(\xi_1)=E(\xi_2)=\cdots=E(\xi_n)=\mu$$

then the expected value can be written as

$$E(\bar{\xi})=\mu \tag{2.21}$$

Again from Eq. 2.20, it is possible to have the variance of the sample mean as

$$\mathrm{Var}(\xi) = \frac{1}{\mathrm{n}^2}[\mathrm{Var}(\xi_1) + \mathrm{Var}(\xi_2) + \cdots + \mathrm{Var}(\xi_\mathrm{n})]$$

Letting the variance of identical populations of random variables $\xi_1, \xi_2, \ldots, \xi_n$ be σ^2, we have,

$$\mathrm{Var}(\xi_1) = \mathrm{Var}(\xi_2) = \cdots = \mathrm{Var}(\xi_\mathrm{n}) = \sigma^2$$

Then the variance of the sample mean becomes

$$\mathrm{Var}(\xi) = \frac{\sigma^2}{\mathrm{n}} \tag{2.22}$$

and the standard deviation of sample mean is $\sigma/\sqrt{n}$.

In case that the samples with a size of n are continuously random-sampled from a specific population to obtain their sample means, then it is almost certain that these sample means would follow a probability distribution with a population mean of $E(\bar{\xi})$ and a population variance of $\mathrm{Var}(\xi)$. Since the formulations of $E(\bar{\xi})$ and $\mathrm{Var}(\xi)$ are deduced in the case of the unknown population distribution, no matter which probability distribution the population follows, as long as the population mean and variance of μ and σ^2 are given, it is inevitable for $E(\bar{\xi})$ and $\mathrm{Var}(\xi)$ to be μ and σ^2/n respectively.

It is worth noting that two concepts of mean and expected value should not be confused. The expected value in Eq. 2.10 is deduced from the mean, but the mean always is not the expected value. The mean generally includes more comprehensive implications. The expected value represents the mean of possible values of a random variable and is meaningful and significant only in the case of large sample size. When the PDF of a random variable is obtained from a large sample, the expected value determined by using Eq. 2.10 is the population mean and a constant, e.g., the population mean of normal distribution is a constant of value μ.

It is usual for the observed values of fatigue life N to be transferred into the logarithm form and then plotted into the histogram to clearly show the ordered change of data. Usually also, it is necessary to find a curve, i.e., experimental frequency curve to fit the histogram for statistical analysis. Although the investigated subjects of the histogram are varied, their experimental frequency curves display some common features as:

(1) The ordinate of curve is always positive;
(2) There is at least one peak on the centre portion of curve;
(3) The two ends of the curve spread out along left and right directions until the ordinate of curve equals or is near to zero;
(4) The area between the curve and the abscissa axis should be equal to 1.

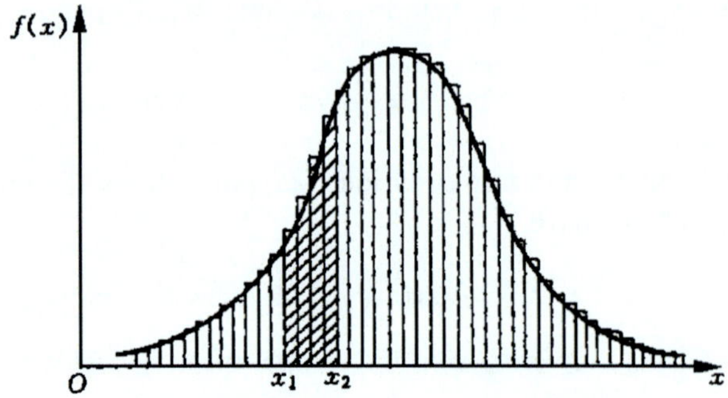

Fig. 2.2 Experimental frequency curve

With increasing observation frequency, the number of grouped data sets increases (shown in Fig. 2.2) and the shape of experimental frequency curve varies less and less until it reaches a stable state. In other words, in the case of $n \to \infty$, the frequency may approximate the probability; the area between frequency curve and horizontal ordinate axis represents the probability; the ordinate of frequency curve depicts probability density and the frequency curve is then called as the probability density curve (PDC). Actually, it is impossible to conduct infinite observations. However, so long as one assumes that infinite individuals exist potentially, no matter whether they are observed one by one or not, the PDC exists objectively and consists of infinite individuals and represents the character of the infinite population. According to the characteristics of the various experimental frequency curves, several probability density functions can be proposed to depict the PDC; the mathematical representation for describing experimental frequency curve is termed as the theoretical frequency function, which is normally known as the PDC in statistics. The normal and Weibull PDFs are usually applied in fatigue reliability.

The normal PDF is denoted as

$$f(x) = \frac{1}{\sigma\sqrt{2\pi}} e^{-\frac{(x-\mu)^2}{2\sigma^2}} \tag{2.23}$$

or

$$f(x) = \frac{1}{\sigma\sqrt{2\pi}} \exp\left[-\frac{(x-\mu)^2}{2\sigma^2}\right] \tag{2.23}$$

where $e = 2.718$ is the base of a natural logarithm. μ and σ are constants. The function of $f(x)$ is the normal PDF. As shown in Fig. 2.3, the normal PDC, i.e., the Gaussian curve, demonstrates the curve is bilaterally symmetric and is suitable for representing the observed value of logarithmic fatigue life.

The Weibull PDF is also suitable for fatigue statistical analysis and is expressed as follows:

$$f(N) = \frac{b}{N_a - N_0}\left[\frac{N - N_0}{N_a - N_0}\right]^{b-1} e^{-\left[\frac{N-N_0}{N_a-N_0}\right]^b} \tag{2.24}$$

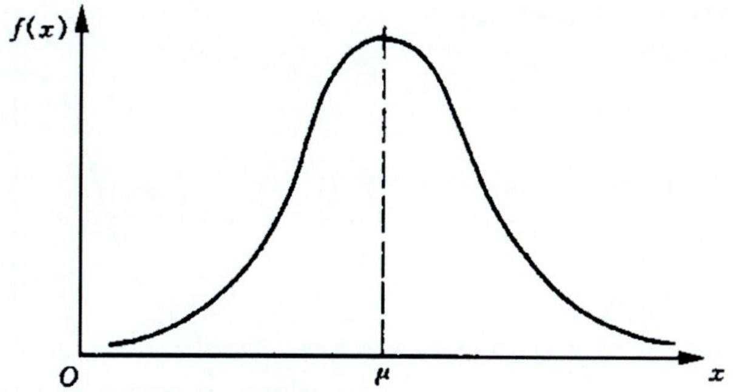

Fig. 2.3 Normal probability density curve

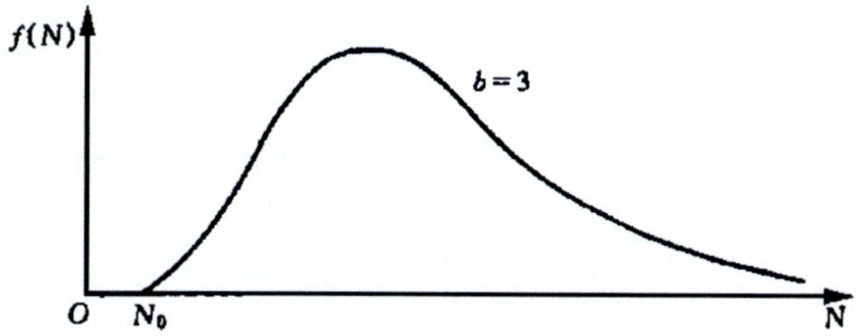

Fig. 2.4 Weibull probability density curve

or

$$f(N) = \frac{b}{N_a - N_0}\left[\frac{N - N_0}{N_a - N_0}\right]^{b-1} \exp\left\{-\left[\frac{N - N_0}{N_a - N_0}\right]^{b}\right\} \quad (2.24)$$

where N_0, N_a and b are three parameters. The Weibull PDC is shown in Fig. 2.4. Figure 2.4 shows that the curve is left–right asymmetric and it intersects the abscissa at N_0. In certain cases, from the actual observed results, the Weibull PDC is seen to be representative of fatigue life N.

In the inference of fatigue life, the statistics U, χ^2 and t, etc., are generally implemented for interval estimation. Assuming a random variable X follows the normal distribution and taking the following transformation:

$$U = \frac{X - \mu}{\sigma} \quad (2.25)$$

then function U is a random variable. Since X samples are in an interval of $(-\infty, \infty)$, with the sample span of $U = (X - \mu)/\sigma$ also being from $-\infty$ to ∞ too, and the PDF of U can then be written as

$$\varphi(u) = \frac{1}{\sqrt{2\pi}} e^{-\frac{u^2}{2}} (-\infty < u < \infty) \quad (2.26)$$

where U is the standard normal variable and $\varphi(u)$ is the standard normal PDF. By comparing Eq. 2.26 with Eq. 2.23, it is found that $\varphi(u)$ is the normal PDF with a population mean of 0 and a standard deviation of 1. Therefore, the standard normal or Gauss distribution is denoted as $N(0;1)$. Equation 2.25 is termed the

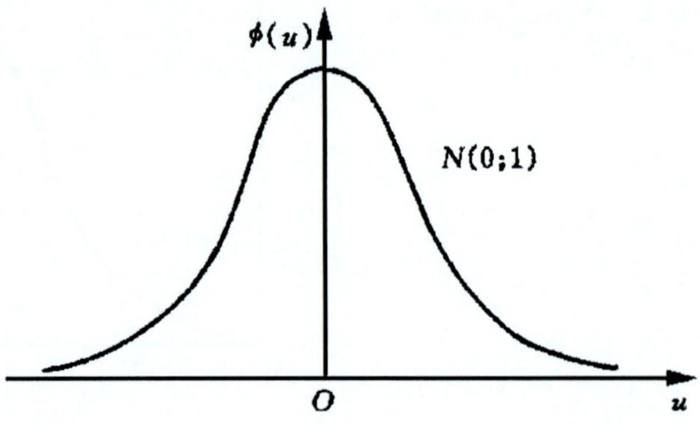

Fig. 2.5 Standard normal probability density curve

standardized substitution of normal variable and the standard normal PDC is shown in Fig. 2.5.

The PDF of random variable χ^2 is expressed as

$$f_v(x) = \frac{\left(\frac{1}{2}\right)^{\frac{v}{2}}}{\Gamma\left(\frac{v}{2}\right)} x^{\frac{v}{2}-1} e^{-\frac{x}{2}} (0 < x < \infty) \tag{2.27}$$

where x is the sampled value of random variable χ^2. v is a parameter of the PDF of χ^2 and is termed as a degree of freedom. With increasing v, the PDC of χ^2 becomes of near symmetric form. The expected value and variance of random variable χ^2 can be derived as:

$$E(\chi^2) = \frac{\alpha}{\beta} = v \tag{2.28}$$

$$\mathrm{Var}(\chi^2) = \frac{\alpha}{\beta^2} = 2v \tag{2.29}$$

The χ^2 distribution shows the following features as:

(1) In case that $U_1, U_2, \ldots, U_v$ are v mutually independent standard normal variables, then $\sum_{i=1}^{v} U_i^2$ follows the χ^2 distribution with degrees of freedom v.
(2) In case where χ_1^2 and χ_2^2 are mutually independent random variables of χ^2 with degrees of freedom v_1 and v_2 respectively, then $\chi_1^2 + \chi_2^2$ is also a random variable following the χ^2 distribution with a degree of freedom $v_1 + v_2$.

   Similarly, it can be deduced that the sum of finite mutually independent random variables of χ^2 is a random variable of χ^2, whose degree of freedom equals the sum of degrees of freedom of all random variables of χ^2.

(3) In the case where s_x^2 represents the variance of a random sample from a normal population $N(\mu;\sigma)$ with a sample size of n, then the random variable $(n-1)s_x^2/\sigma^2$ follows the χ^2 distribution with a freedom degree of $v = n - 1$.

$$\chi^2 = \frac{(n-1)s_x^2}{\sigma^2} \tag{2.30}$$

From the χ^2 distribution, it is possible to have the following random variable function:

$$\eta = \sqrt{\frac{\chi^2}{v}} \tag{2.31}$$

As the sampled span of χ^2 is from 0 to ∞, $\eta = \sqrt{\frac{\chi^2}{v}}$ samples from 0 to ∞ too. Thus the PDF of random variable function of η is obtained as

$$g(y) = \frac{2\left(\frac{v}{2}\right)^{\frac{v}{2}}}{\Gamma\left(\frac{v}{2}\right)} y^{v-1} e^{-\frac{1}{2}vy^2} (0<y<\infty) \tag{2.32}$$

If standard normal variables U and η are mutually independent, then the ratio between these two random variables is known as the t distribution with t_x as the variable:

$$t_x = \frac{U}{\eta} = \frac{U}{\sqrt{\frac{\chi^2}{v}}} \tag{2.33}$$

The sampled span of random variable t_x is from $-\infty$ to ∞. Assuming t_0 be a sampled value of random variable t_x, then the distribution function $P(t_x < t_0)$ becomes

$$P(t_x<t_0) = \int_{-\infty}^{t_0} h(t)dt \tag{2.34}$$

with

$$h(t) = \frac{\Gamma\left(\frac{v+1}{2}\right)}{\sqrt{\pi v}\Gamma\left(\frac{v}{2}\right)}\left(1 + \frac{t^2}{v}\right)^{-\frac{v+1}{2}} \tag{2.35}$$

where $h(t)$ is the PDF of t. Because $h(t)$ is an even function, the t PDC is similar to the standard normal PDC and is bilaterally symmetric relative to the ordinate axis. Further mathematical proof can be deployed to demonstrate that in the case of $v \to \infty$, the t distribution is nearly the same as the standard normal distribution; in reality, in the case of $v \geq 30$, both distributions are very close.

2.3 Probability Distribution of Fatigue Life

As mentioned above, the normal and Weibull PDFs are usually applied in reliability analysis of fatigue life. The normal PDC expressed by Eq. 2.23 is shown in Fig. 2.6. From Fig. 2.6, it is clear that the curve has a maximum of $f(x)$, a symmetry axis at an abscissa value of μ and two inflexions located on the curve at $x = \mu \pm \sigma$. Bilaterally symmetric sections of the curve spread out along an asymptote to the abscissa. The shape of the curve depends on the population standard deviation σ. The larger the σ, flatter is the shape of curve. This implies greater dispersion. In turn, less the σ, the sharper is the shape of curve and less is the dispersion. In the case that μ and σ are known, the normal PDC can be determined completely. The simple notation $N(\mu;\sigma)$ is implemented to expediently denote the normal distribution with the population mean and standard deviation of μ and σ separately.

For a specific normal PDF, the distribution function $F(x_p)$ of the normal variable, i.e., the probability of the normal variable X being less than a value of x_p can be obtained as:

$$F(x_p) = P(X < x_p) = \frac{1}{\sigma\sqrt{2\pi}} \int_{-\infty}^{x_p} e^{-\frac{(x-\mu)^2}{2\sigma^2}} dx \tag{2.36}$$

where $F(x_p) = P(X < x_p)$, in geometrical terms, implies the area between the curve from $-\infty$ to x_p and the abscissa axis, i.e., the dashed area as shown in Fig. 2.7. If logarithmic fatigue life follows the normal distribution, then $F(x_p)$ equals the rate of failure. In the case of a known normal PDF, Eq. 2.36 shows that the value of $F(x_p)$ is fully dependent on x_p. With the ordinate and abscissa being $F(x_p)$ and x_p, the distribution function curve can be drawn, see Fig. 2.8. It is seen that $F(x_p)$ increases with increasing of x_p; this is owed to the increase in the dashed area between the curve to the left of x_p and the abscissa (shown in Fig. 2.7). In the case of $x_p = \mu$, the dashed area should be 0.5, or $F(x_p) = 0.5$, whereas in the case of x_p approaching $-\infty$ or ∞, the limits of $F(x_p)$ are 0 and 1 respectively.

From the normal PDF, it is possible to have the cumulative frequency function, i.e., the probability of the normal variable X being greater than a value of x_p:

$$P(X > x_p) = \frac{1}{\sigma\sqrt{2\pi}} \int_{x_p}^{\infty} e^{-\frac{(x-\mu)^2}{2\sigma^2}} dx \tag{2.37}$$

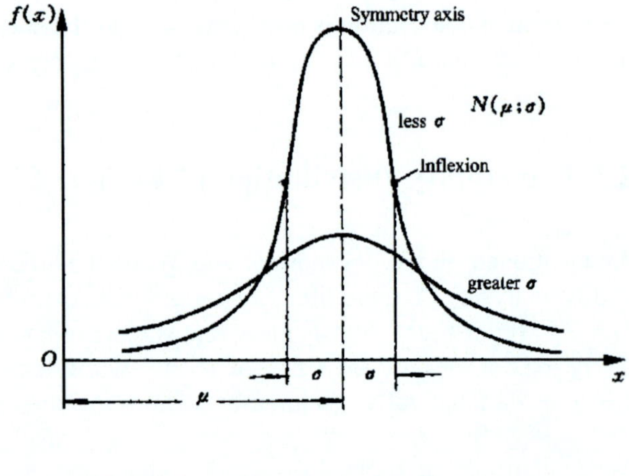

Fig. 2.6 Normal probability density curve

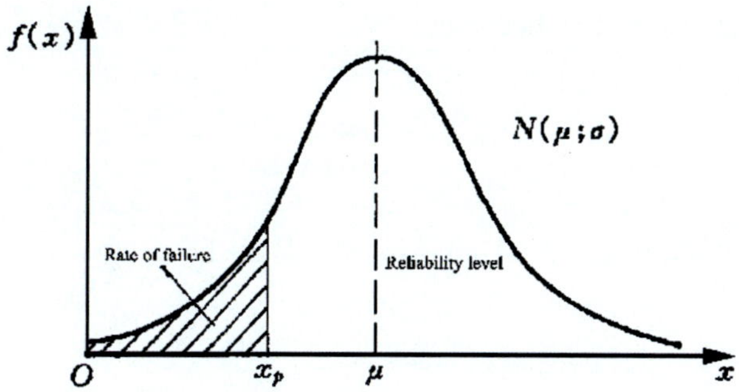

Fig. 2.7 Normal probability density curve

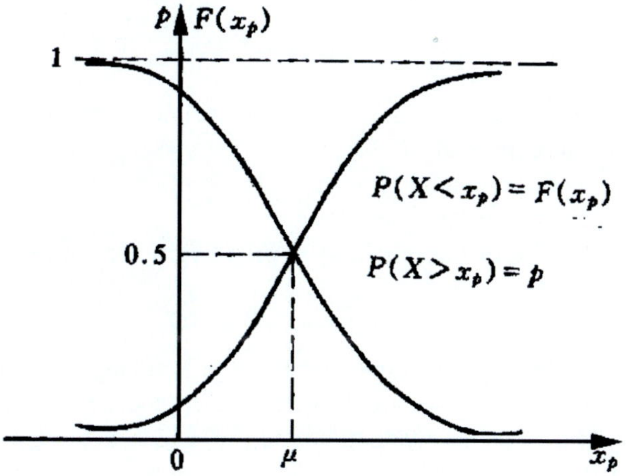

Fig. 2.8 Distribution function and cumulative frequency curves

If x represents the logarithmic fatigue life, the cumulative frequency function, i.e., $P(X > x_p)$ is equivalent to the reliability level p and is also a function of x_p, which has the following relationship with $P(X < x_p)$ as:

$$P(X > x_p) + P(X < x_p) = 1 \tag{2.38}$$

The cumulative frequency curve is shown in Fig. 2.8 too. From Fig. 2.8, it is evident that $P(X > x_p)$ decreases with increase in x_p, while the reverse is true with $P(X < x_p)$ increasing with decrease in x_p.

The cumulative frequency function $P(X > x_p)$ of normal variable plays an important role in fatigue reliability analysis. The standardized substitution method of variable is employed to integrate Eq. 2.37; let

$$u = \frac{x - \mu}{\sigma} \tag{2.39}$$

$$du = \frac{dx}{\sigma}, dx = \sigma du$$

then Eq. 2.37 becomes

$$P(X > x_p) = \frac{1}{\sigma\sqrt{2\pi}} \int_{x_p}^{\infty} e^{-\frac{(x-\mu)^2}{2\sigma^2}} dx = \frac{1}{\sqrt{2\pi}} \int_{u_p}^{\infty} e^{-\frac{u^2}{2}} du \tag{2.40}$$

From Eq. 2.39, the lower limit of the integral is

$$u_p = \frac{x_p - \mu}{\sigma} \tag{2.41}$$

Equation 2.40 shows that the integrand function is transferred to be a standard normal PDF as:

$$\varphi(u) = \frac{1}{\sqrt{2\pi}} e^{-\frac{u^2}{2}}$$

Hence, $P(X > x_p)$ can be represented by the area between the normal (or the standard normal) PDC and the abscissa. u_p is called as the standard normal deviator pertaining to a reliability level of p. Obviously, the relationship between x_p and p is established through u_p.

In case of the reliability level $p = 50\%$, then $u_p = 0$ and

$$x_{50} = \mu \tag{2.42}$$

Evidently, the population mean of μ equals to logarithmic fatigue life pertinent to a reliability level of 50%. Fatigue life N_{50} pertaining to a reliability level of 50% is termed as the median fatigue life, which is the antilogarithm of x_{50}. N_{50} means that the lives of half the individuals among the population are greater than N_{50}, whereas those of other half are less than N_{50}.

If X_1 and X_2 are statistically independent normal variables with the population means of μ_1 and μ_2 and population standard deviations of σ_1 and σ_2 respectively, then from Eqs. 2.13 and 2.17, it can be show that $\varsigma = X_1 + X_2$ inevitably becomes a normal variable too with an expected value of $\mu_1 + \mu_2$, a variance of $\sigma_1^2 + \sigma_2^2$ together with a standard deviation of $\sqrt{\sigma_1^2 + \sigma_2^2}$. The PDF of ς is

$$f(x) = \frac{1}{\sqrt{\sigma_1^2 + \sigma_2^2}\sqrt{2\pi}} \exp\left\{ -\frac{[x - (\mu_1 + \mu_2)]^2}{2(\sigma_1^2 + \sigma_2^2)} \right\} \tag{2.43}$$

Using the same method as in the above derivation, it can be deduced that the difference between two normal variables of X_1 and X_2, $\varsigma = X_1 - X_2$, is also the normal variable with a mathematical expectation of $\mu_1 - \mu_2$ and a variance of $\sigma_1^2 + \sigma_2^2$. The above findings on the sum or difference between two normal variables can be extended to the case of n normal variables of $X_1, X_2, \ldots, X_n$ with the population means of $\mu_1, \mu_2, \ldots, \mu_n$ and the population standard deviations of $\sigma_1, \sigma_2, \ldots, \sigma_n$ respectively. Letting

$$\varsigma = a_1X_1 + a_2X_2 + \cdots + a_nX_n \tag{2.44}$$

then ς is a normal variable with the following expected value:

$$E(\varsigma) = a_1\mu_1 + a_2\mu_2 + \cdots + a_n\mu_n \tag{2.45}$$

and the variance:

$$\mathrm{Var}(\varsigma) = a_1^2\sigma_1^2 + a_2^2\sigma_2^2 + \cdots + a_n^2\sigma_n^2 \tag{2.46}$$

Therefore, it can be concluded that the homogeneous linear function ς of statistically independent normal variables is also a normal variable with the expected value and variance determined by using Eqs. 2.45 and 2.46 respectively.

If fatigue life is denoted as N, assuming a random variable $X = \log N$, then the PDF of logarithmic fatigue life follows the normal distribution, or

$$f(x) = \frac{1}{\sigma\sqrt{2\pi}}\exp\left[-\frac{(x-\mu)^2}{2\sigma^2}\right]$$

From the above equation, it is possible to have the PDF of fatigue life as

$$p(N) = \frac{1}{\sigma N\sqrt{2\pi}\ln 10}e^{-\frac{(\log N-\mu)^2}{2\sigma^2}} \quad (0<N<\infty) \tag{2.47}$$

It is worth noticing that μ and σ in Eq. 2.47 are the population mean and standard deviation of logarithmic fatigue life respectively. The PDC of fatigue life N is shown in Fig. 2.9. From Eq. 2.47, one has the expected value, i.e., population mean μ_N of fatigue life N as

$$\mu_N = \exp\left[\frac{1}{2}\sigma^2\ln^2 10 + \mu\ln 10\right]$$

or

$$\log\mu_N = \frac{\ln\mu_N}{\ln 10} = \frac{1}{2}\sigma^2\ln 10 + \mu \tag{2.48}$$

Equation 2.48 reveals the relationship between the population mean and variance of μ and σ^2 of normal variable X and the population mean μ_N of random

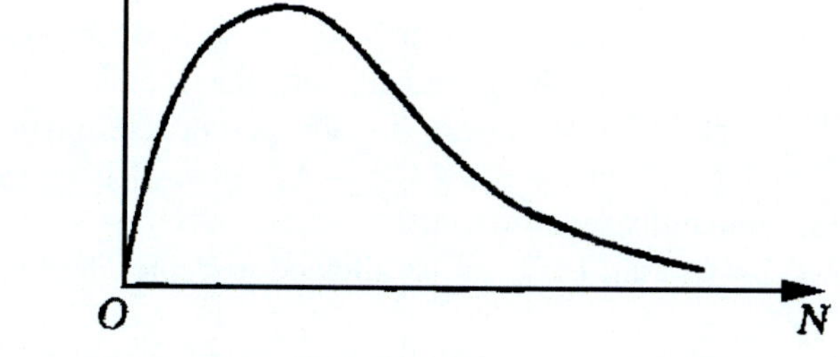

Fig. 2.9 Probability density curve of fatigue life

variable N. μ is regarded as the population mean of logarithmic fatigue life ($x = \log N$). If the sample size is denoted as n, then the estimator of μ is

$$\hat{\mu} = \frac{1}{n}(\log N_1 + \log N_2 + \cdots + \log N_n)$$

For the normal distribution, $\hat{\mu}$ is equivalent to the estimator of logarithmic fatigue life pertaining to a reliability level of 50%, alternatively

$$\hat{\mu} = \log \hat{N}_{50}$$

where $\hat{N}_{50}$ is the estimator of fatigue life pertaining to a reliability level of 50% (i.e., median estimator of fatigue life) and equals the geometric mean of fatigue lives observed at values of N_1, N_2, ..., N_n as

$$\hat{N}_{50} = (N_1 N_2, \ldots, N_n)^{\frac{1}{n}}$$

However, μ_N represents the population mean of fatigue life N, whose estimator is the arithmetic mean of fatigue lives as

$$\hat{\mu}_N = \frac{1}{n}(N_1 + N_2 + \cdots + N_n)$$

By comparing the above two estimators of $\hat{\mu}$ and $\hat{\mu}_N$, it is easy to find the differences between μ and μ_N. Equation 2.48 shows that in the case of the logarithmic fatigue life following the normal distribution, the difference between $\log \mu_N$ and μ is $\frac{1}{2}\sigma^2 \ln 10$, and the population variance of logarithmic fatigue life is σ^2. Furthermore, it can be proved that the reliability level p corresponding to logarithmic fatigue life of $\log N_p$ is the reliability level p_N pertaining to fatigue life N_p. Moreover, from Eq. 2.47, it is possible to obtain the reliability level p_N pertinent to a specific fatigue life N_p as [3]

$$p_N = \frac{1}{\sigma\sqrt{2\pi}\ln 10} \int_{N_p}^{\infty} \frac{1}{N} e^{-\frac{(\log N - \mu)^2}{2\sigma^2}} dN \tag{2.49}$$

As mentioned above, the normal distribution is suitable for the cases of medium and short life ranges, whereas the Weibull distribution fits better for fatigue life in long life range of greater than 10^6 cycles. The Weibull PDF has the advantage of a minimum safe life, i.e., the safe life corresponding to a reliability level of 100%, while from the normal distribution theorem, only in the case where the logarithmic safe life $x_p = \log N_p$ is near $-\infty$, or $N_p = 0$, the reliability level is 100%. Evidently, this is in disagreement with the actual case. In order to overcome this drawback, it is necessary to add an undetermined parameter N_0 to replace $x_p = \log N_p$ with $x_p = \log (N_p - N_0)$; here N_0 is the minimum safe life pertaining to a reliability level of 100%.

The Weibull PDF can be allowed to depict the distribution law of fatigue life N as

$$f(N) = \frac{b}{N_a - N_0}\left[\frac{N - N_0}{N_a - N_0}\right]^{b-1} \exp\left\{-\left[\frac{N - N_0}{N_a - N_0}\right]^{b}\right\}(N_0 < N < \infty) \quad (2.50)$$

where N_0 is the minimum life parameter, N_a is the characteristic life parameter and b is the Weibull shape parameter (slope parameter). Due to the Weibull PDF being characterised by three parameters unlike the normal distribution having only two, i.e., μ and σ, the Weibull PDF may more perfectly fit the experiments than the normal distribution.

In the case of $b = 1, f(N)$ in Eq. 2.50 becomes a simple exponential PDF. In the case of $b = 2$, $f(N)$ is the Rayleigh PDF and in the case of $b = 3 \sim 4, f(N)$ is close to the normal PDF. The Weibull PDC is shown in Fig. 2.10. Figure 2.10 shows that the peak of curve always deviates to the left and the deviation varies with the change of b. For $b > 1$, the curve intersects the abscissa at $N = N_0$ and exists a high-positive minimum life of N_0, the difference of $(N_a - N_0)$ is greater, the curve becomes flatter and the dispersity is larger. The right end of curve spreads out to infinity along an asymptote to the abscissa. Furthermore, as shown below, it can be proved that like other PDFs, the Weibull PDF satisfies the condition of $\int_{N_0}^{\infty} f(N)dN = 1$, i.e., the area between the curve and the horizontal ordinate axis equals to 1.

If $b = 2$, then the Weibull distribution becomes the Rayleigh PDF as

$$f(N) = \frac{2N}{N_a^2} e^{-\frac{N^2}{N_a^2}}$$

The random variable following the Weibull distribution, i.e., the Weibull variable, is denoted as N_ξ. From Eq. 2.50, one has the distribution function $F(N_p)$ of the Weibull variable, namely, the probability $P(N_\xi < N_p)$ of N_ξ being less than a value of N_p as

$$F\left(N_p\right) = P\left(N_\xi < N_p\right) = \int_{N_0}^{N_p} f(N)dN \quad (2.51)$$

Equation 2.51 represents the area between the curve from N_0 to N_p and the abscissa, as the dashed area shown in Fig. 2.11. Substituting Eq. 2.50 into Eq. 2.51 yields

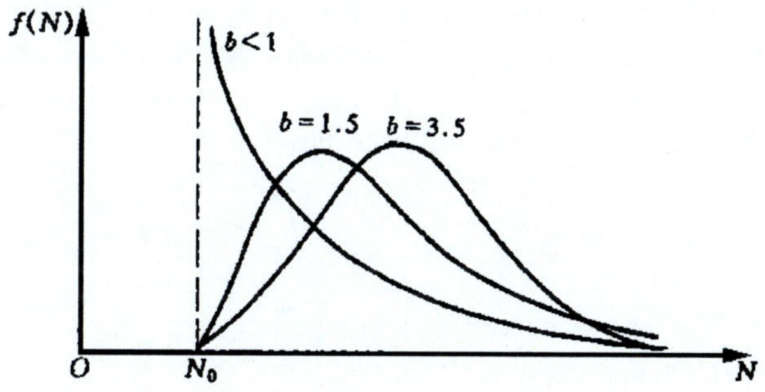

Fig. 2.10 Weibull probability density curve

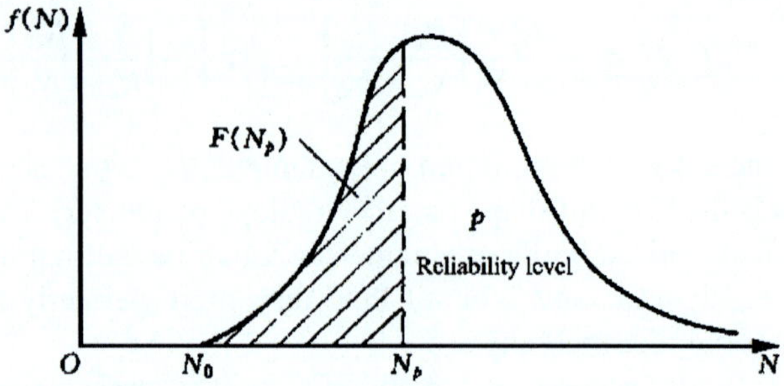

Fig. 2.11 Weibull probability density curve

$$F(N_p) = \int_{N_0}^{N_p} \frac{b}{N_a - N_0}\left[\frac{N - N_0}{N_a - N_0}\right]^{b-1} \exp\left\{-\left[\frac{N - N_0}{N_a - N_0}\right]^b\right\} dN \tag{2.52}$$

Taking the following transformation

$$\left[\frac{N - N_0}{N_a - N_0}\right]^b = Z, \frac{N - N_0}{N_a - N_0} = Z^{\frac{1}{b}}, \frac{dN}{N_a - N_0} = \frac{1}{b} Z^{\frac{1-b}{b}} dZ$$

then in the case of $N = N_0$, $Z = 0$, and in the case of $N = N_p$, $Z = \left[\frac{N_p - N_0}{N_a - N_0}\right]^b$.

Taking the integral transformation of Z with a lower limit of 0 and an upper limit of $Z_p = \left[\frac{N_p - N_0}{N_a - N_0}\right]^b$, then Eq. 2.52 becomes

$$F(N_p) = \int_0^{Z_p} \frac{b}{N_a - N_0} Z^{b-1} e^{-Z} \frac{N_a - N_0}{b} Z^{\frac{1-b}{b}} dZ = \int_0^{Z_p} e^{-Z} dZ = -\left[e^{-Z}\right]_0^{Z_p} = 1 - e^{-Z_p}$$

Substituting Z_p into the above equation, one has the distribution function as

$$F(N_p) = 1 - \exp\left\{-\left[\frac{N_p - N_0}{N_a - N_0}\right]^b\right\} \tag{2.53}$$

The distribution function curve is shown in Fig. 2.12 with an ordinate of $P(N_\xi < N_p) = F(N_p)$ and an abscissa of N_p. It can be observed that $P(N_\xi < N_p)$ increases with increasing of N_p, this is because the area between the curve from N_0 to N_p and the abscissa increases with the shifting of N_p to the right (shown in Fig. 2.11). Equation 2.53 reveals that in the case of $N_p \to \infty$, the limit of $P(N_\xi < N_p)$ is 1 (shown in Fig. 2.12). Substituting $P(N_\xi < N_p) = 1$ and $N_p = \infty$ into Eq. 2.51, it is possible to have

$$\int_{N_0}^{\infty} f(N) dN = 1$$

The above equation demonstrates that the area between the curve and the horizontal ordinate axis equals unity.

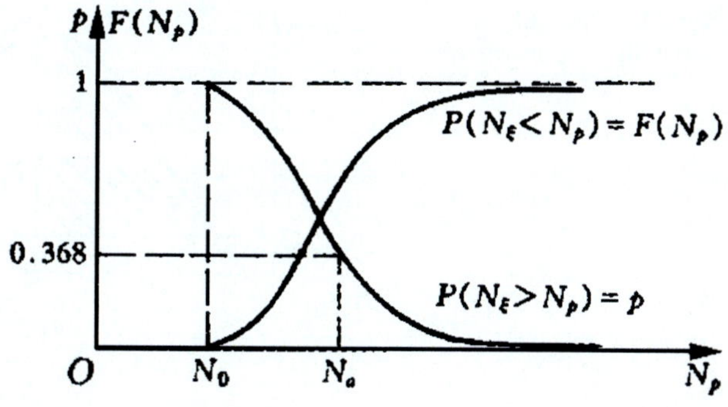

Fig. 2.12 Distribution function and cumulative frequency curves

The distribution function $P(N_\xi < N_p)$ is equivalent to the rate of failure and cumulative frequency function $P(N_\xi > N_p)$ equals the reliability level, or

$$P\left(N_\xi > N_p\right) = 1 - P\left(N_\xi < N_p\right) = \exp\left\{-\left[\frac{N_p - N_0}{N_a - N_0}\right]^b\right\} \tag{2.54}$$

The cumulative frequency function $P(N_\xi > N_p)$ is denoted as the reliability level p as

$$p = \exp\left\{-\left[\frac{N_p - N_0}{N_a - N_0}\right]^b\right\} \tag{2.55}$$

In the case of known parameters of N_0, N_a and b together with a specific reliability level p, from Eq. 2.55, one has a safe life of N_p, i.e., fatigue life pertinent to a reliability level of p. The curve of $p = P\left(N_\xi > N_p\right)$ is shown in Fig. 2.12 too. From Fig. 2.12, it is clear that when $N_p = N_0$, $p = 1$, that is the minimum life N_0 is the safe life pertaining to a reliability level of 100%. When $N_p = N_a$, from Eq. 2.55, it is possible to have

$$p = \exp\left\{-\left[\frac{N_p - N_0}{N_a - N_0}\right]^b\right\} = e^{-1} = \frac{1}{2.718} = 36.8\%$$

This implies that the characteristic life parameter N_a is fatigue life corresponding to a reliability level of 36.8% (shown in Fig. 2.12). Owing to the Weibull PDF requiring three parameters of N_0, N_a and b, but without both parameters of μ and σ^2, so it is necessary to employ the three parameters of N_0, N_a and b to derive the parameters of μ and σ^2.

From Eq. 2.10, the definition of mathematical expectation of the Weibull variable N_ξ can be written as

$$E(N_\xi) = \int_{N_0}^{\infty} Nf(N)dN$$

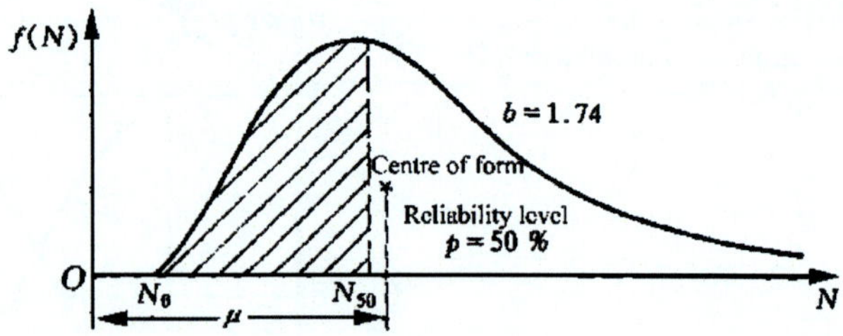

Fig. 2.13 Weibull probability density curve

Substituting Eq. 2.50 into the above equation and taking the following transformation of variable as

$$\left[\frac{N-N_0}{N_a-N_0}\right]^b = Z, \frac{N-N_0}{N_a-N_0} = Z^{\frac{1}{b}}, \frac{dN}{N_a-N_0} = \frac{1}{b}Z^{\frac{1-b}{b}}dZ \tag{2.56}$$

then

$$E(N_\xi) = \int_0^\infty b\left(Z^{\frac{1}{b}} + \frac{N_0}{N_a-N_0}\right)Z^{\frac{b-1}{b}}e^{-Z}\frac{(N_a-N_0)}{b}Z^{\frac{1-b}{b}}dZ$$
$$= (N_a-N_0)\int_0^\infty Z^{\left(1+\frac{1}{b}\right)-1}e^{-Z}dZ + N_0$$

From the definition of $\Gamma(\alpha)$ function, the integral item in the above equation becomes

$$\int_0^\infty Z^{\left(1+\frac{1}{b}\right)-1}e^{-Z}dZ = \Gamma\left(1+\frac{1}{b}\right)$$

Hence, it is possible to have the mathematical expectation of the Weibull population mean μ as a function of the three parameters that define this distribution as

$$\mu = E(N_\xi) = N_0 + (N_a - N_0)\Gamma\left(1+\frac{1}{b}\right) \tag{2.57}$$

According to the geometric meaning of mathematical expectation, the population mean μ is the centre position of form of the area between the PDC $f(N)$ and the abscissa (shown in Fig. 2.13), while the population median N_{50} represents fatigue life N_p corresponding to a reliability level of 50%. From Fig. 2.13, it can be observed that in the case of $b = 1.74$, the peak of curve deviates to the left and $\mu > N_{50}$, whereas for the normal population, because of the symmetry of the curve, both population mean and median are concurrent as demonstrated in Eq. 2.51: $\mu = N_{50}$. Therefore, it is essential to implement the median fatigue life or strength, and not by the mean, to obtain fatigue behaviour of material. If the mean equals to the median, i.e., $\mu = N_{50}$, then the Weibull PDF is close to the normal PDF and the shape parameter of the Weibull distribution $b = 3.57$.

From Eq. 2.12, it is possible to derive the variance $\mathrm{Var}(N_\xi)$ of the Weibull variable as

$$\sigma^2 = \mathrm{Var}(N_\xi) = (N_a - N_0)^2\left[\Gamma\left(1+\frac{2}{b}\right) - \Gamma^2\left(1+\frac{1}{b}\right)\right] \tag{2.58}$$

Equation 2.58 is regarded as a measure of the population dispersion. Equation 2.58 shows that σ^2 increases with increasing $(N_a - N_0)$ and decreases with increasing b.

The Weibull distribution has a strong compatibility and flexibility to fit the experimental data, since the shape of PDC is capable of deviating to the left and right with the deviation being determined through a skew coefficient.

2.4 Point Estimation of Population Parameter

Estimating the population parameters, e.g., μ and σ^2 from a sample is termed as point estimation and a sample with sample size greater than 50 is regarded as a large sample. In the tests of fatigue life, only one value can be determined from one specimen. In many circumstances, due to time and resource constraints, it is infeasible to conduct extensive experimental investigations in order to generate large numbers of datasets required by classical statistical processes. In contrast, only small numbers of sample data (sample size $n < 50$) can be provided.

When the sample eigenvalues are taken as the estimators of population parameter, generally, it is necessary to satisfy the demands of consistency and unbiasedness. In the case of sample size $n \to \infty$, the sample mean $\bar{x}$ becomes the expected value $E(\xi)$ of random variable and the population distribution coincides with the distribution of random variable. This is because the sample mean $\bar{x}$ approximates uniformly to the population mean μ. Similarly, in the case of sample size $n \to \infty$, the sample variance s^2 approximates uniformly to the random variable variance $\mathrm{Var}(\xi)$, i.e., population variance σ^2. Obviously, in case where the sample mean $\bar{x}$ and variance s^2 are regarded as the estimators $\hat{\mu}$ and $\hat{\sigma}^2$ of population mean μ and variance σ^2 respectively, then the estimators become closer to the truths of population parameter with increasing of sample size n.

Unbiased estimator means that the expected value of estimator as a random variable determined from each sample with a sampled size of n should equal the estimated population parameter. For example, if the sample mean $\bar{x}$ is taken as the unbiased estimator $\hat{\mu}$ of population mean μ, then it is necessary for the expected value of sample mean to be equal to μ. Thus, the unbiased estimator $\hat{\mu}$ of population mean should satisfy the following condition

$$E(\hat{\mu}) = \mu \tag{2.59}$$

As a random variable, the sample mean may be written as

$$\bar{\xi} = \frac{1}{n}\sum_{i=1}^{n} \xi_i$$

From Eq. 2.21, the expected value of $\bar{\xi}$ just equals to μ, or

$$E(\bar{\xi}) = \mu$$

The above equation reveals that the sample mean $\bar{x}$ satisfies the unbiasedness condition as the estimator of population mean μ, so

$$\bar{x} = \hat{\mu} \tag{2.60}$$

Letting the observed values of fatigue life to be N_1, N_2, ..., N_n, then the estimator of normal population mean of logarithmic fatigue life is

$$\hat{\mu} = \bar{x} = \frac{1}{n}\sum_{i=1}^{n} \log N_i$$

For the normal distribution, it is possible to have

$$\mu = x_{50} = \log N_{50}$$

From the above two equations, one has the estimator $\hat{N}_{50}$ of median fatigue life as

$$\log \hat{N}_{50} = \frac{1}{n}\sum_{i=1}^{n} \log N_i$$

or

$$\hat{N}_{50} = (N_1 N_2 \ldots N_n)^{\frac{1}{n}}$$

In case that fatigue life follows the Weibull distribution, then the population mean μ of the Weibull distribution can also be estimated by using the sample mean $\bar{N}$ as

$$\hat{\mu} = \bar{N} = \frac{1}{n}\sum_{i=1}^{n} N_i \tag{2.61}$$

Similarly, the unbiased estimator $\hat{\sigma}^2$ of population variance should satisfy the following condition as

$$E(\hat{\sigma}^2) = \sigma^2 \tag{2.62}$$

The sample variance s^2 satisfies the condition (2.62) as the unbiased estimator $\hat{\sigma}^2$ of population variance σ^2, thus

$$s^2 = \hat{\sigma}^2 \tag{2.63}$$

Table 2.1 Correction coefficient $\hat{k}$ of standard deviation

n	5	6	7	8	9	10	11	12	13	14
$\hat{k}$	1.063	1.051	1.042	1.036	1.031	1.028	1.025	1.023	1.021	1.020
n	15	16	17	18	19	20	30	40	50	60
$\hat{k}$	1.018	1.017	1.016	1.015	1.014	1.014	1.009	1.006	1.005	1.005

No matter which distribution the population follows, Eqs. 2.60 and 2.62 are suitable for the estimation of mean and variance parameters. Note that the estimators are not equivalent to the truths of population parameters of μ and σ^2 absolutely. Only in the case of large enough sampling, is the estimator near the truth. The unbiased estimator of population variance is

$$\hat{\sigma}^2 = s^2 = \frac{\sum_{i=1}^{n}(x_i - \bar{x})^2}{n-1} = \frac{\sum_{i=1}^{n} x_i^2 - \frac{1}{n}\left(\sum_{i=1}^{n} x_i\right)^2}{n-1}$$

The estimator of standard deviation is obtained by the square root of $\hat{\sigma}^2$, i.e., $\hat{\sigma} = s$. Strictly speaking, the sample standard deviation s is a biased estimator of population standard deviation since the unbiased condition of $E(s_\xi) = \sigma$ is not satisfied. In fatigue reliability design, the sample standard deviation s is usually corrected to find an unbiased estimator of population standard deviation to remove the bias by using χ^2 distribution. However, such unbiased estimator fits only for the normal population. The unbiased estimator of normal population standard deviation can be written as

$$\hat{\sigma} = \hat{k}s \tag{2.64}$$

where

$$\hat{k} = \sqrt{\frac{n-1}{2}}\frac{\Gamma\left(\frac{n-1}{2}\right)}{\Gamma\left(\frac{n}{2}\right)} \tag{2.65}$$

$\hat{k}$ is the coefficient of correction of standard deviation. If the population follows the normal distribution, the unbiased estimator of population standard deviation can be obtained from Eq. 2.64. The coefficients of correction corresponding to different sample size are listed in Table 2.1. Table 2.1 shows that there is a small difference between the correction coefficient $\hat{k}$ and 1; as a result, no correction is conducted in general engineering application and $\hat{\sigma} = s$ is taken. This is especially so in the case of $n > 50$; then $\hat{k} \to 1$. Thereby, for large sample sizes, the sample standard deviation s always is the unbiased estimator of population standard deviation σ. However, in fatigue reliability design of aeronautics and marine structural parts, it is desirable to correct the sample standard deviation s. Generally, the unbiased estimator $\hat{\mu} = \bar{x}$ of population mean μ is suitable for no matter which distribution the population follows. Consequently, for the normal population, the population mean μ is the population median and the sample mean is the estimator of population median. In the case where the population follows the normal distribution, substituting $\hat{\mu} = \bar{x}$ and $\hat{\sigma} = \hat{k}s$ into Eq. 2.41, it is possible to have the estimator of a percentile x_p pertaining to a reliability level of p as

$$\hat{x}_p = \hat{\mu} + u_p\hat{\sigma} = \bar{x} + u_p\hat{k}s \tag{2.66}$$

Thus the estimators of each fatigue life N_i pertinent to a reliability level could be determined from small number of samples. A sample with a sample size of n is random-sampled from a known population to obtain n observed values for arranging in sequential queue from smaller to greater as

$$x_1 < x_2 < \cdots < x_i < \cdots < x_n$$

where i is the arranged ordinal of the observed value from smaller to greater. If the PDF of the population is denoted as $f(x)$, then the rate of failure $F(x_i)$ (distribution function) of ith observed value x_i can be determined. No matter which distribution the random-sampled population follows, or which PDF $f(x)$ is, the mathematical expectation of the rate of failure corresponding to x_i is $i/(n + 1)$, which is termed as the mean rank and regarded as the estimator of population failure rate in engineering. Therefore, the estimator of population reliability level p pertaining to ith observed value x_i becomes

$$\hat{p} = 1 - \frac{i}{n+1} \tag{2.67}$$

In case of only one specimen for fatigue test, i.e., $n = 1$, then from Eq. 2.67, the estimator of reliability level pertinent to fatigue life of specimen is only 50%, namely

$$\hat{p} = 1 - \frac{i}{n+1} = 1 - \frac{1}{1+1} = 50\%$$

2.5 Interval Estimation of Population Mean and Standard Deviation

In reality, no true population parameters, e.g., mean μ and standard deviation σ, etc., are known. And it is hard for the point estimator of population parameter determined from a sample with a finite sample size to amount to the theoretical truth obtained from infinite observed values. As a consequence, sometimes it is feasible to use an interval limit for estimating population parameter to indicate the error of estimation. At a specific probability, the location interval of population parameter can be estimated by using the sample eigenvalues and is called as the interval estimation of population parameter. Though the estimator $\bar{x}$ of population mean μ satisfies the consistency and unbiasedness demands, it is possible for the

Fig. 2.14 Standard normal probability density curve

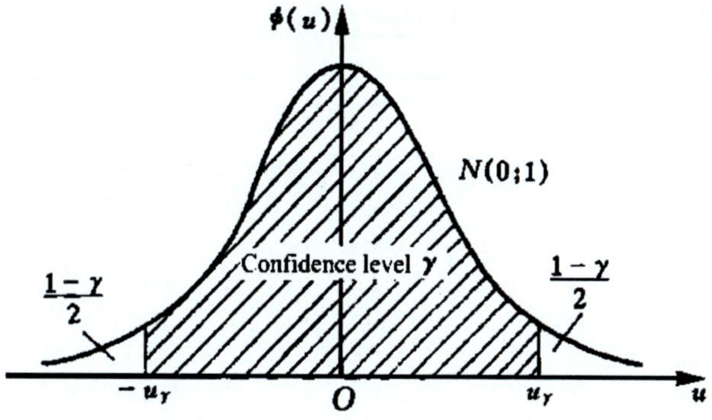

sample mean $\bar{x}$ determined from a small sample with finite observed values to be close to, but impossible to equal to the population mean μ. Therefore, it is imprecise for finite observed values to be applied to estimate the population mean, whereas it is feasible for the sample mean to be employed for estimating the location interval population mean pertaining to a specific probability, which is termed as the interval estimation of population mean.

In the interval estimation of population mean μ, assuming μ be unknown, then the standard normal variable can be written as

$$u = \frac{\bar{x} - \mu}{\sigma_0/\sqrt{n}}$$

In general, it is possible to select a probability of γ from the area between the standard normal PDC and the abscissa as the dashed area shown in Fig. 2.14. Thus the blank areas between two ends of curve and the abscissa are $(1-\gamma)/2$ respectively; the corresponding u_γ can be determined from the numerical tabular representation of the normal distribution (listed in Table 2.2). So the standard normal variable locates in an interval of $(-u_\gamma;\ u_\gamma)$ at a probability of γ, that is

$$-u_\gamma < u < u_\gamma$$

where is termed as the confidence level. We then have

$$-u_\gamma < \frac{\bar{x} - \mu}{\sigma_0/\sqrt{n}} < u_\gamma$$

or

$$\bar{x} - u_\gamma \frac{\sigma_0}{\sqrt{n}} < \mu < \bar{x} + u_\gamma \frac{\sigma_0}{\sqrt{n}} \tag{2.68}$$

Equation 2.68 is the interval estimation formula of the normal population mean μ. Equation 2.68 demonstrates that the confidence level of the interval $\left(\bar{x} - u_\gamma\sigma_0/\sqrt{n}; \bar{x} + u_\gamma\sigma_0/\sqrt{n}\right)$ including the population mean μ equals to γ, here the interval is called as the confidence interval, and $\bar{x} + u_\gamma\sigma_0/\sqrt{n}$ and $\bar{x} - u_\gamma\sigma_0/\sqrt{n}$ are termed as the confidence upper and lower limits respectively. The pre-condition

Table 2.2 Numerical tabular of and u

u	γ	u	γ	u	γ
3.0	0.9974	1.5	0.8664	4.753	0.999998
2.9	0.9962	1.4	0.8384	4.265	0.99998
2.8	0.9948	1.3	0.8064	3.719	0.9998
2.7	0.9930	1.2	0.7698	3.090	0.998
2.6	0.9906	1.1	0.7286	2.576	0.990
2.5	0.9876	1.0	0.6824	2.326	0.980
2.4	0.9836	0.9	0.6318	1.960	0.950
2.3	0.9786	0.8	0.5762	1.645	0.900
2.2	0.9722	0.7	0.5160	1.282	0.800
2.1	0.9642	0.6	0.4514	1.036	0.700
2.0	0.9544	0.5	0.3830	0.842	0.600
1.9	0.9426	0.4	0.3108	0.674	0.500
1.8	0.9282	0.3	0.2358	0.524	0.400
1.7	0.9108	0.2	0.1586	0.385	0.300
1.6	0.8904	0.1	0.0796	0.253	0.200
				0.126	0.100
				0	0.000

to apply Eq. 2.68 is to know the population standard deviation, or to deal with large sample. However, sometimes it is hard to satisfy the above conditions. Actually, it is feasible to apply the t distribution theorem to treat a sample with a sample size of greater than 5 from the practical experience. Assuming $\bar{X}$ denotes the sample mean with a sample size of n random-sampled from the normal population $N(\mu;\sigma)$, then the standard normal variable becomes

$$U = \frac{\bar{X} - \mu}{\frac{\sigma}{\sqrt{n}}} \tag{2.69}$$

Substituting Eq. 2.69 into Eq. 2.33 yields

$$t_x = \frac{\frac{\bar{X}-\mu}{\sigma}\sqrt{n}}{\sqrt{\frac{\chi^2}{v}}} = \frac{(\bar{X} - \mu)\sqrt{vn}}{\sigma\sqrt{\chi^2}}$$

If s_x^2 denotes the sample variance randomly-sampled from the normal population $N(\mu;\sigma)$, then the freedom degree of $\chi^2 = (n-1)s_x^2/\sigma^2$ is $v = n - 1$. Substituting χ^2 and v into the above equation gives the t_x variable with a freedom degree of $v = n - 1$ as

$$t_x = \frac{\bar{X} - \mu}{s_x}\sqrt{n} \tag{2.70}$$

A confidence level of γ is selected to determine two abscissa values of t and $-t_\gamma$ (shown in Fig. 2.15), between which the area below the curve (i.e., the dashed area in Fig. 2.15) amounts to γ. Thus, the t_γ is obtained as

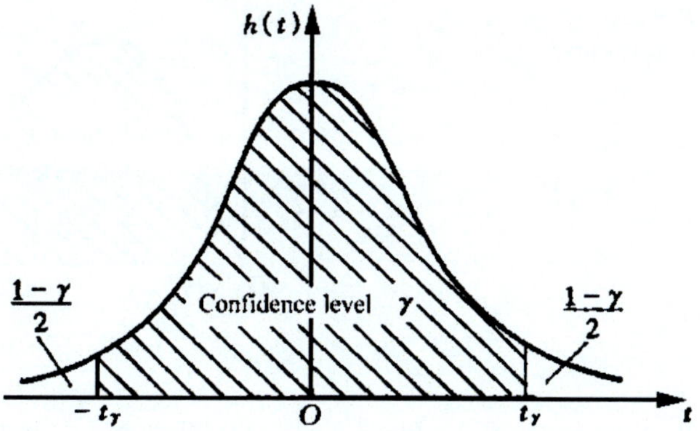

Fig. 2.15 t probability density curve

$$\int_{t_\gamma}^{\infty} h(t)dt = \frac{1-\gamma}{2} \tag{2.71}$$

Since the probability of the t_x variable located in an interval of $(-t_\gamma,\ t_\gamma)$ is equivalent to γ (i.e., a confidence level of γ), it is possible to have the following inequality

$$-t_\gamma < t_x < t_\gamma \tag{2.72}$$

Substituting Eq. 2.70 into Eq. 2.72 shows

$$-t_\gamma < \frac{\bar{X}-\mu}{s_x}\sqrt{n} < t_\gamma$$

where μ is assumed to be an undetermined value. Assuming random variables $\bar{X}$ and s_x be sampled as $\bar{x}$ and s respectively in a sampling, then the above in equation becomes

$$-t_\gamma < \frac{\bar{x}-\mu}{s}\sqrt{n} < t_\gamma \tag{2.73}$$

Equation 2.73 can also be written as

$$\bar{x} - t_\gamma\frac{s}{\sqrt{n}} < \mu < \bar{x} + t_\gamma\frac{s}{\sqrt{n}} \tag{2.74}$$

Equation 2.74 is the interval estimation formula of the normal population mean μ, demonstrating that the confidence level of the confidence interval $\left(\bar{x} - t_\gamma\frac{s}{\sqrt{n}};\bar{x} + t_\gamma\frac{s}{\sqrt{n}}\right)$ including the population mean μ equals γ.

From Fig. 2.15, it is observed that greater the confidence level γ, greater is the value of $|t_\gamma|$, and wider is the confidence interval. In fact, it is desirable for the confidence interval to be less and for the confidence level to be greater. However, the above theorem reveals that less the confidence interval, less the confidence

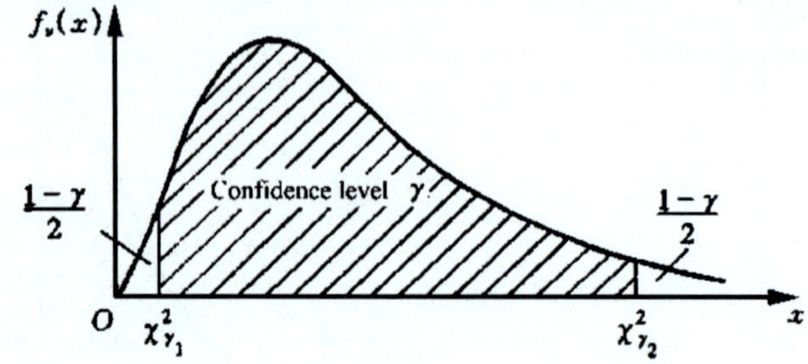

Fig. 2.16 χ^2 probability density curve

level becomes. To address this contradiction, i.e., not only to decrease the confidence interval but also to keep the confidence level high, it is necessary to increase the sample size of n to reduce the value of $t_\gamma \frac{s}{\sqrt{n}}$ and then to lower the confidence interval of $\left(\bar{x} - t_\gamma \frac{s}{\sqrt{n}}; \bar{x} + t_\gamma \frac{s}{\sqrt{n}}\right)$.

The transformation form of Eq. 2.74 becomes

$$-\frac{st_\gamma}{\bar{x}\sqrt{n}} < \frac{\mu - \bar{x}}{\bar{x}} < \frac{st_\gamma}{\bar{x}\sqrt{n}} \tag{2.75}$$

where $(\mu - \bar{x})/\bar{x}$ is the relative error of sample mean $\bar{x}$ to population mean μ. The relative error limit (absolute value) is denoted as δ, or

$$\delta = \frac{st_\gamma}{\bar{x}\sqrt{n}} \tag{2.76}$$

where δ is a small quantity in a span from 1 to 10%. In the case where $\bar{x}$, s and n satisfy Eq. 2.76, Eq. 2.75 demonstrates that the confidence level of the relative error of sample mean to population median being less than $\pm\delta$ is equal to γ. Hence, by using Eq. 2.76, the least number of observed values (or effective specimens) is obtained.

From Eq. 2.30, it is possible to conduct the interval estimation of population standard deviation. Similarly, a confidence level of γ is selected to determine an interval of $(\chi^2_{\gamma_1}, \chi^2_{\gamma_2})$ (shown in Fig. 2.16), between which the area between the curve and the abscissa (i.e., the dashed area in Fig. 2.16) amounts to γ. In the case that γ and v are known, from Table 2.3 of χ^2 distribution. $\chi^2_{\gamma_1}$ and $\chi^2_{\gamma_2}$ are obtained to make the probability of the χ^2 variable locating in the interval of $(\chi^2_{\gamma_1};\chi^2_{\gamma_2})$ to equal to γ, alternatively, for a confidence level of γ, one has

$$\chi^2_{\gamma_1} < \chi^2 < \chi^2_{\gamma_2}$$

Substituting Eq. 2.30 into the above equation yields

$$\chi^2_{\gamma_1} < \frac{(n-1)s_x^2}{\sigma^2} < \chi^2_{\gamma_2}$$

Table 2.3 Numerical tabular of χ_γ^2

γ / υ	0.0	0.80	0.90	0.95	0.98	0.99
3	2.366	6.251	7.815	9.348	11.345	12.838
4	3.357	7.779	9.488	11.143	13.277	14.860
5	4.351	9.236	11.070	12.823	15.068	16.750
6	5.348	10.645	12.592	14.449	16.812	18.548
7	6.346	12.017	14.067	16.013	18.475	20.278
8	7.344	13.362	15.507	17.535	20.090	21.955
9	8.343	14.684	16.919	19.023	21.666	23.589
10	9.342	15.987	18.307	20.483	23.209	25.188
11	10.341	17.275	19.675	21.920	24.725	26.757
12	11.340	18.549	21.026	23.337	26.217	28.300
13	12.340	19.812	22.362	24.736	27.688	29.819
14	13.339	21.064	23.685	26.119	29.141	31.319
15	14.339	22.307	24.996	27.488	30.578	32.801
16	15.338	23.542	26.296	28.845	32.000	34.267
17	16.338	24.769	27.587	30.191	33.409	35.718
18	17.338	25.989	28.869	31.526	34.805	37.156
19	18.338	27.204	30.144	32.852	36.191	38.582
20	19.337	28.412	31.410	34.170	37.566	39.997
21	20.337	29.615	32.671	35.479	38.932	41.401
22	21.337	30.813	33.924	36.781	40.289	42.796
23	22.337	32.007	35.172	38.076	41.638	44.181
24	23.337	33.196	36.415	39.364	42.80	45.558
25	24.337	34.382	37.652	40.646	44.314	46.928

Assuming σ be an unknown value and random variable s_x be sampled as s in a sampling, then the above inequality becomes

$$\chi^2_{\gamma_1} < \frac{(n-1)s^2}{\sigma^2} < \chi^2_{\gamma_2}$$

Inverting the above inequality leads to

$$\frac{1}{\chi^2_{\gamma_2}} < \frac{\sigma^2}{(n-1)s^2} < \frac{1}{\chi^2_{\gamma_1}}$$

or

$$s\sqrt{\frac{n-1}{\chi^2_{\gamma_2}}} < \sigma < s\sqrt{\frac{n-1}{\chi^2_{\gamma_1}}} \tag{2.77}$$

Equation 2.77 is the interval estimation formula of normal population standard deviation σ, demonstrating that the confidence level of the confidence interval $\left(s\sqrt{(n-1)/\chi^2_{\gamma_2}}\,;\, s\sqrt{(n-1)/\chi^2_{\gamma_1}}\right)$ including the value of σ is γ.

2.6 Interval Estimation of Population Percentile

The normal population percentile x_p is defined in probability form as

$$P(X > x_p) = \int_{x_p}^{\infty} f(x)dx = p$$

where $f(x)$ is the PDF of normal variable X. From Eq. (2.41), the value of $x_p = \mu + u_p\sigma$ corresponding to a reliability level of p is the population percentile, which represents the logarithmic safe life or safe fatigue strength. From Eq. 2.66, the percentile estimator is

$$\hat{x}_p = \bar{x} + u_p\hat{\sigma}$$

where $\hat{\sigma} = \hat{k}s$. It is hard for the sample percentile $(\bar{x} + u_p\hat{\sigma})$ to just equal to the population true value $(\mu + u_p\sigma)$; in other words, it is possible for the sample percentile $(\bar{x} + u_p\hat{\sigma})$ to be greater or less than the population truth $(\mu + u_p\sigma)$. Thus there is an error between the predictions for sample percentile $\bar{x} + u_p\hat{\sigma}$ of normal random variable X, determined from Eq. 2.66, and the population true value $\mu + u_p\sigma$. Also the confidence level of statistical results estimated from small samples needs to be analysed. The confidence level is a statistical variable correlated with sample size. Usually, the confidence levels are fixed and the amount of data determines the width of the corresponding confidence intervals, i.e., the confidence interval decreases with increase in the amount of data (shown in Fig. 2.17). While the width of the corresponding confidence intervals is fixed and the amount of data determines the confidence levels, e.g., under a fixed width of corresponding confidence intervals, the fewer the samples, the less data items are available and thus the lower the confidence level in the statistical results. With larger sample numbers, there are more data items and thus the confidence level in the statistical results is higher. The sample percentile $\bar{x} + u_p\hat{\sigma}$ of normal random variable X pertinent to reliability level p, estimated using a large sample size, is near the population true value and the confidence levels are generally high. Thus it

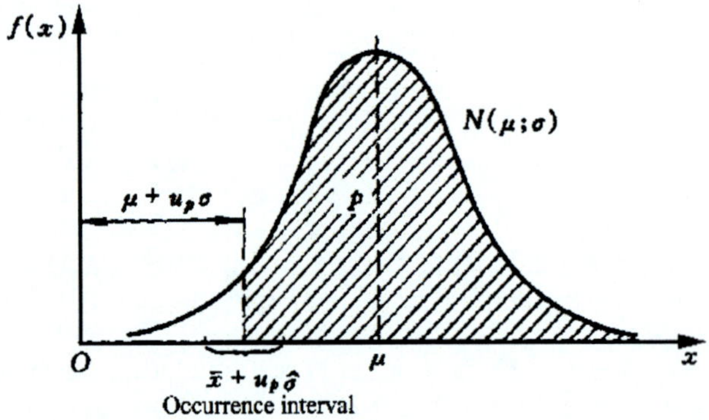

Fig. 2.17 Normal probability density curve

is not necessary for the confidence level of large sample sets to be analyzed; only the reliability level need be estimated using Eq. 2.66. A log-normal distribution lends itself to a perfect theoretical solution to estimate the unbiased distribution parameter values, the confidence level and the confidence interval. It is thus an apt approach for fatigue life estimation and is widely applied in reliability-based design approaches using small sample numbers. Using the log-normal distribution and Eq. 2.66, the confidence level of safe fatigue life can be determined easily. In addition, it is also possible to predict the minimum number of specimens required for fatigue tests for a given confidence level.

As is well known, the t-distribution is generally used for the calculation of confidence interval for the estimator of mean value and variance of a random variable. However, a theoretical solution of the confidence interval for the estimator of sample percentile of a random variable does not currently exist. Consequently, it is essential to establish the relationship between t-statistics and sample percentile $\bar{x} + u_p\hat{\sigma}$ of the normal random variable X and to derive a theoretical solution to determine the confidence interval of sample percentile $\bar{x} + u_p\hat{\sigma}$ of normal random variable X pertinent to reliability level p from small sample numbers.

From Eqs. 2.65 and 2.66, the sample percentile $\bar{x} + u_p\hat{\sigma}$ can be written into the random variable function as

$$\zeta = \bar{X} + u_p\hat{k}s_x$$

where $\bar{X}$ and s_x denote the random variables of sample mean and sample standard deviation respectively. In practical application, ζ is assumed to approximately follow the normal distribution. Therefore it is possible to calculate the expected value $E(\zeta)$ and variance $\mathrm{Var}(\zeta)$.

$$E(\zeta) = E\left(\bar{X} + u_p\hat{k}s_x\right) = E(\bar{X}) + u_pE\left(\hat{k}s_x\right)$$

From Eqs. 2.60 and 2.64, it can be show that $E(\bar{X}) = \mu$ and $\hat{\sigma} = \hat{k}s$ is the unbiased estimator of normal population standard deviation, i.e.,

$$E\left(\hat{k}s_x\right) = \sigma$$

Thus, $E(\zeta)$ becomes

$$E(\zeta) = \mu + u_p\sigma \tag{2.78}$$

The variance $\mathrm{Var}(\zeta)$ can be written as

$$\mathrm{Var}(\zeta) = \mathrm{Var}\left(\bar{X} + u_p\hat{k}s_x\right)$$

or

$$\mathrm{Var}(\zeta) = \mathrm{Var}(\bar{X}) + u_p^2\hat{k}^2\mathrm{Var}(s_x) \tag{2.79}$$

From Eq. 2.22, one has

$$\mathrm{Var}(\bar{X}) = \frac{\sigma^2}{n} \tag{2.80}$$

Again, from Eq. 2.30, the χ^2 variable with degree of freedom $v = n - 1$ is

$$\frac{(n-1)s_x^2}{\sigma^2} = \chi^2 \tag{2.81}$$

$$s_x = \frac{\sigma}{\sqrt{n-1}}\chi \tag{2.82}$$

Then,

$$\mathrm{Var}(s_x) = \mathrm{Var}\left(\frac{\sigma}{\sqrt{n-1}}\chi\right) = \frac{\sigma^2}{n-1} Var(\chi) \tag{2.83}$$

From Eqs. 2.12 and 2.83, it can be shown that

$$\mathrm{Var}(s_x) = \frac{\sigma^2}{n-1}\left\{E(\chi^2) - [E(\chi)]^2\right\} \tag{2.84}$$

Based on Eq. 2.28, expected value of the χ^2 variable is

$$E(\chi^2) = \int_0^{\infty} x f_v(x) dx = v$$

By means of the statistics, it is also possible to have

$$E(\chi) = \sqrt{2}\frac{\Gamma\left(\frac{v+1}{2}\right)}{\Gamma\left(\frac{v}{2}\right)}$$

Substituting the above two equations into Eq. 2.84 shows

$$\mathrm{Var}(s_x) = \frac{\sigma^2}{n-1}\left\{n - 1 - 2\left[\frac{\Gamma\left(\frac{n}{2}\right)}{\Gamma\left(\frac{n-1}{2}\right)}\right]^2\right\} \tag{2.85}$$

Again, substituting Eqs. 2.80 and 2.85 into Eq. 2.79 yields

$$\mathrm{Var}(\zeta) = \frac{\sigma^2}{n} + \frac{u_p^2 \hat{k}^2 \sigma^2}{n-1}\left\{n - 1 - 2\left[\frac{\Gamma\left(\frac{n}{2}\right)}{\Gamma\left(\frac{n-1}{2}\right)}\right]^2\right\} \tag{2.86}$$

By using Eq. 2.65, the simplified form of above equation is deduced as

$$\mathrm{Var}(\zeta) = \sigma^2\left[\frac{1}{n} + u_p^2(\hat{k}^2 - 1)\right] \tag{2.87}$$

From Eqs. 2.78 and 2.87, the standard normal variable is derived as

$$U = \frac{\zeta - E(\zeta)}{\sqrt{\mathrm{Var}(\zeta)}} = \frac{\left(\bar{X} + u_p\hat{k}s_x\right) - \left(\mu + u_p\sigma\right)}{\sigma\sqrt{\frac{1}{n} + u_p^2\left(\hat{k}^2 - 1\right)}} \tag{2.88}$$

In terms of Eq. 2.33, the t_x variable is

$$t_x = \frac{U}{\sqrt{\frac{\chi^2}{v}}}$$

Substituting Eqs. 2.81 and 2.88 as well as $v = n - 1$ into the above equation allows

$$t_x = \frac{\left(\bar{X} + u_p\hat{k}s_x\right) - \left(\mu + u_p\sigma\right)}{s_x\sqrt{\frac{1}{n} + u_p^2\left(\hat{k}^2 - 1\right)}}$$

Thus, in a sampling, the sampled value of t_x variable is

$$t = \frac{\left(\bar{x} + u_p\hat{\sigma}\right) - \left(\mu + u_p\sigma\right)}{s\sqrt{\frac{1}{n} + u_p^2\left(\hat{k}^2 - 1\right)}} \tag{2.89}$$

In the case of a specific confidence level of γ and degree of freedom of $v = n - 1$, the t_γ value can be obtained in an interval of $(-t_\gamma, t_\gamma)$ pertaining to a confidence level of γ, namely,

$$-t_\gamma < t < t_\gamma$$

or

$$-t_\gamma < \frac{\left(\bar{x} + u_p\hat{\sigma}\right) - \left(\mu + u_p\sigma\right)}{s\sqrt{\frac{1}{n} + u_p^2\left(\hat{k}^2 - 1\right)}} < t_\gamma$$

Transforming the above inequality yields

$$-\frac{t_\gamma s\sqrt{\frac{1}{n} + u_p^2\left(\hat{k}^2 - 1\right)}}{\bar{x} + u_p\hat{\sigma}} < \frac{\left(\bar{x} + u_p\hat{\sigma}\right) - \left(\mu + u_p\sigma\right)}{\bar{x} + u_p\hat{\sigma}} < \frac{t_\gamma s\sqrt{\frac{1}{n} + u_p^2\left(\hat{k}^2 - 1\right)}}{\bar{x} + u_p\hat{\sigma}} \tag{2.90}$$

Equation 2.90 is the interval estimation formula of the normal population percentile. The minimum number of observed values (or specimens) can be determined by stipulating that no error between $(\bar{x} + u_p\hat{\sigma})$ and $(\mu + u_p\sigma)$ exceeds a limit of δ pertinent to a specific confidence level of γ. The error limit is denoted as δ, then

$$\delta = \frac{t_\gamma s\sqrt{\frac{1}{n} + u_p^2\left(\hat{k}^2 - 1\right)}}{\bar{x} + u_p\hat{\sigma}}$$

Since $\hat{\sigma} = \hat{k}s$, the function of error limit δ with respect to the coefficient of variation $s/\bar{x}$ is

$$\delta = \frac{t_\gamma\left(\frac{s}{\bar{x}}\right)\sqrt{\frac{1}{n} + u_p^2\left(\hat{k}^2 - 1\right)}}{1 + u_p\hat{k}\left(\frac{s}{\bar{x}}\right)} \tag{2.91}$$

For a reliability level of $p = 50\%$, $u_p = 0$ and Eq. 2.91 degenerates into Eq. 2.76. The error limit δ generally is selected from 1 to 10%. Thus it is possible to establish the minimum number of specimens to estimate the population percentile as

$$\frac{s}{\bar{x}} \leq \frac{\delta}{t_\gamma\sqrt{\frac{1}{n} + u_p^2\left(\hat{k}^2 - 1\right)} - 0.05u_p\hat{k}} \tag{2.92}$$

where $s/\bar{x}$ is calculated from n observed values. In the case where n satisfies Eq. 2.92 and there is no system error, the confidence level of the relative error of $(\bar{x} + u_p\hat{\sigma})$ to the population truth $(\mu + u_p\sigma)$ being less than $\pm\delta$, amounts to γ.

It is worth pointing out that the reliability level and confidence level are two different concepts, e.g., the confidence level of $\gamma = 95\%$ implies that among 100 estimators of logarithmic safe life $(\bar{x} + u_p\hat{\sigma})$ pertinent to a reliability level of p, determined from 100 samples, the relative error of 95 estimated values to the truth $(\mu + u_p\sigma)$ is less than $\pm5\%$. Obviously, the confidence level is proposed with regard to the sample, whereas the reliability level is defined regarding the individual.

According to the t distribution theorem, one has the lower confidence limit of logarithmic fatigue life corresponding to a confidence level of γ, with

$$P\left\{t < t_\gamma\right\} = \gamma \tag{2.93}$$

where t_γ is the γ percentile of t distribution corresponding to a confidence level γ. Substituting Eqs. 2.89 and 2.64 into Eq. 2.93 and taking transformation gives

$$P\left\{\left(\bar{x} + u_p\hat{k}s\right) - t_\gamma s\sqrt{\frac{1}{n} + u_p^2\left(\hat{k}^2 - 1\right)} < \left(\mu + u_p\sigma\right)\right\} = \gamma \tag{2.94}$$

From Eq. 2.94, it is possible to have the one-sided lower confidence limit of logarithmic fatigue life pertaining to a reliability level of p and a confidence level of γ as [4]

$$\hat{x}_{p\gamma} = \log \hat{N}_{p\gamma} = \left(\bar{x} + u_p\hat{k}s\right) - t_\gamma s\sqrt{\frac{1}{n} + u_p^2\left(\hat{k}^2 - 1\right)} \tag{2.95}$$

Similarly, the one-sided upper confidence limit of logarithmic fatigue life pertinent to a reliability level of p and a confidence level of γ is

$$\hat{x}_{p\gamma} = \log \hat{N}_{p\gamma} = \left(\bar{x} + u_p\hat{k}s\right) + t_\gamma s\sqrt{\frac{1}{n} + u_p^2\left(\hat{k}^2 - 1\right)} \tag{2.96}$$

References

1. Gao ZT (1981) Statistics applied in fatigue. National Defense Industry Press, Beijing
2. Gao ZT, Xiong JJ (2000) Fatigue Reliability. Beihang University Press, Beijing
3. Xiong JJ, Gao ZT (1997) The probability distribution of fatigue damage and the statistical moment of fatigue life. Sci China (Ser E) 40(3):279–284
4. Xiong J, Shenoi RA, Gao Z (2002) Small sample theory for reliability design. J Strain Anal Eng Des 37(1):87–92

Chapter 3
Principles Underpinning Reliability based Prediction of Fatigue and Fracture Behaviours

3.1 Introduction

Because of the stochastic nature of the damage processes, it is generally inadvisable to formulate a deterministic approach to predict fatigue damage and fatigue life. Thus, many stochastic mathematical expressions for fatigue life and fatigue damage process have been developed. Comprehensive reviews of this subject have been conducted by Weibull [1], Freudenthal [2], Bogdanoff [3], Lin [4], Provan [5] and Sobczyk [6], et al. Due to variations between individual specimens, fatigue data can be described by random variables to study the variability of damage and life and to analyse their average trends [1, 2]. With improvements in crack-size measurements, fatigue crack growth data can be described in a random time–space and state-space to depict local variations within a single specimen analyse. This has been done by a stationary lognormal process-based randomized approach of deterministic crack growth equation in power law and polynomial forms [7–10]. Under random spectrum loading, a fatigue cumulative process can be described by discrete Markov chain models [3, 6] or continuous Markov process methods based on the solution of the Fokker–Planck equation [4, 6, 11, 12]. The appropriateness and accuracy of these categories of stochastic models have been verified by statistically meaningful fatigue crack growth data sets with certain degrees of accuracy [7, 8, 13–15]. In fact, all above categories of stochastic models have been obtained by randomizing the deterministic fatigue crack growth equation and using large sample numbers of statistically meaningful data sets from which to determine the random coefficients for the models.

However, in many circumstances, due to time and resource constraints, it is infeasible to conduct extensive experimental investigations in order to generate the large numbers of data sets required by classical statistical processes. Hence, it is desirable to have a technique to address the paucity of data in assessing the structural probability of fatigue and fracture performance. In the chapter, a series of original and practical approaches of deterministic equation are proposed for the

J. J. Xiong and R. A. Shenoi, *Fatigue and Fracture Reliability Engineering*,
Springer Series in Reliability Engineering, DOI: 10.1007/978-0-85729-218-6_3,

assessment of structural probability fatigue and fracture performance from a minimal dataset [16–19].

3.2 A Randomized Approach to a Deterministic Equation

As mentioned in Sect. 1.6, the simplest model that describes crack growth rate under complex spectrum loading is a power law relation [20]:

$$\frac{da}{dt} = C(K_{\max})^m \tag{3.1}$$

with

$$K_{\max} = S\sqrt{\pi \cdot a}Y(a) \tag{3.2}$$

where C and m are the undetermined material constants. S is the fatigue stress. $Y(a)$ is the stress intensity boundary correction factor. Over relatively large crack length intervals it is possible to approximate the stress intensity factor by a power series of the crack length a as

$$K_{\max} = S\sum_{i=1}^{\infty} c_i a^{b_i} \tag{3.3}$$

where c_i and b_i are constants. In a region of small crack sizes, Eq. 3.3 can be implemented with a further approximation to retain only the first term and substituted into Eq. 3.1. Then one has

$$\frac{da}{dt} = C\left(Sc_1 a^{b_1}\right)^m = Qa^b \tag{3.4}$$

where $Q = C(Sc_1)^m$ and $b = b_1 m$ are parameters which depend on loading spectrum, material properties and structural geometry. In literature [21–23], Eq. 3.4 has been verified to be reasonably accurate and has been applied to describe the crack growth rate for spectrum loading particularly for small cracks. Separating the variables in Eq. 3.4 and integrating, the fatigue crack propagation a–t curve under complex spectrum loading can be expressed as

$$t - \tau_0 = Ca^m \tag{3.5}$$

with

$$m = 1 - b \tag{3.6}$$

$$\tau_0 = t_0 - \frac{a_0^{1-b}}{Q(1-b)} \tag{3.7}$$

$$C = \frac{1}{Q(1-b)} \tag{3.8}$$

where τ_0 is the undetermined material constant and t_0 is the initial time.

In order to account for the random nature of crack propagation, in general, random factors, which may be a random variable, a random process of time, or a random process of space, have been introduced and added as a random disturbance to the deterministic model. From this, a random differential equation can be obtained. The solution of the randomized differential equation is probabilistic in nature as well. The integral expression of fatigue crack growth and randomized model of Eq. 3.5 can be written as

$$t - \tau_0 = W(a)Ca^m \tag{3.9}$$

where the random factor, $W(a)$ is a non-negative, stationary lognormal random process of space with a median value of 1.0 and a standard deviation σ_w. The natural logarithmic form of Eq. 3.9 can be expressed as

$$\ln(t - \tau_0) = \ln C + m \cdot \ln a + \ln W(a) \tag{3.10}$$

Let

$$Z(a) = \ln W(a) \tag{3.11}$$

Then Eq. 3.10 becomes

$$\ln(t - \tau_0) = \ln C + m \cdot \ln a + Z(a) \tag{3.12}$$

and one can have $Z(a)$ as a normal random process with zero mean and standard deviation

$$\sigma_z = \sqrt{\ln\left(1 + \sigma_w^2\right)} \tag{3.13}$$

A general auto-correlation function $R_{zz}(\hat{a})$ of the following form is assumed to depend only on the crack size difference $\hat{a}$ for the random process $W(a)$

$$R_{zz}(\hat{a}) = \sigma_w^2 \exp(-\varsigma|\hat{a}|) \tag{3.14}$$

where ς^{-1} indicates a measure of the correlation distance for $W(a)$ and will be called 'correlation space' hereafter for simplicity. The reason for using the above exponentially decaying auto-correlation function lies in the viewpoint that the auto-correlation function of the crack growth rate should decrease as the crack size difference $\hat{a}$ increases. By selecting appropriate values of ς, different degrees of fatigue scatter can be fitted using the proposed stochastic fatigue crack growth model. The Fourier transform of the autocorrelation function, denoted by $\Phi_{zz}(\omega)$, is referred to as the power spectral density [4].

$$\Phi_{zz}(\omega) = \frac{1}{2\pi}\int_{-\infty}^{\infty} R_{zz}(\hat{a})e^{-i\omega\hat{a}}d\hat{a} = \frac{2\varsigma}{\varsigma^2 + \omega^2}\sigma_w^2 \tag{3.15}$$

where $i = \sqrt{-1}$ and ω is the frequency in radians per second. Within the class of random process $Z(a)$ or $W(a)$, two extreme cases should be considered, because of mathematical simplicity:

i) when $\varsigma \rightarrow \infty$, the autocorrelation function $R_{zz}(\hat{a})$ becomes a Dirac delta function:

$$R_{zz}(\hat{a}) = \sigma_w^2 \delta(\hat{a}) \tag{3.16}$$

and

$$\Phi_{zz}(\omega) = \sigma_w^2 / 2\pi \tag{3.17}$$

The random process $Z(a)$ or $W(a)$ is totally uncorrelated at any two crack size instants. Such a random process is referred to as the white noise process.

ii) when $\varsigma \rightarrow 0$, the autocorrelation function $R_{zz}(\hat{a})$ becomes a constant, i.e.:

$$R_{zz}(\hat{a}) = \sigma_w^2 \tag{3.18}$$

or the random process $Z(a)$ or $W(a)$ is totally correlated at any two time instants. Hence $Z(a)$ or $W(a)$ becomes a random variable and the crack propagation model is referred to as the lognormal random variable model.

In reality, the stochastic behavior of crack propagation lies between the two extreme cases described above. Nevertheless, it is worth pointing out that

a) Using Monte Carlo method, the general lognormal random process model can be applied to simulate the stochastic crack propagation. However, there is still no practical method to determine the measure of the correlation distance ς^{-1} from large sample number of statistically meaningful data sets [5, 13, 15].
b) The lognormal white noise process results in very little statistical dispersion for crack propagation and hence it is not a valid model [5, 7].
c) The random variable model can give a good prediction of the mean value and a larger statistical dispersion for crack growth damage distribution than the experimental data and hence it is conservative. Yang and Donath [24] found that the random variable model is adequate to describe stochastic crack growth subjected to spectral loading.

Therefore, it is possible to implement the random variable model to analyse the statistical properties of Eq. 3.12. Again, letting

$$Y = \ln(t - t_0), b_1 = \ln C, b_2 = m, X = \ln a \tag{3.19}$$

then Eq. 3.18 becomes

$$Y = b_1 + b_2 X + Z(a) \tag{3.20}$$

From Eq. 3.20, it is apparent that for a specific crack size a, random process Y follows a normal random variable with mean value $b_1 + b_2 x$ and standard deviation σ_z. Thus, the probability density function of random variable Y can be given by

$$f(y) = \frac{1}{\sigma_z \cdot \sqrt{2\pi}} \cdot \exp\left[-\frac{1}{2\sigma_z^2}(y - b_1 - b_2 x)^2\right] \tag{3.21}$$

and the likelihood function of random variable Y is

$$L(b_1, b_2, \tau_0, \sigma_z) = \prod_{i=1}^{n} \left\{\frac{1}{\sigma_z \sqrt{2\pi}} \cdot \exp\left[-\frac{1}{2\sigma_z^2}(y_i - b_1 - b_2 x_i)^2\right]\right\} \tag{3.22}$$

where n is the sample size. The natural logarithm form of above likelihood function can be written as

$$\ln L = -n \cdot \ln \sigma_z - \frac{n}{2} \cdot \ln(2\pi) - \frac{1}{2\sigma_z^2} \cdot \sum_{i=1}^{n} (y_i - b_1 - b_2 x_i)^2 \tag{3.23}$$

From Eq. 3.23, according to the maximum likelihood principle, which is a method used for fitting a statistical model to data, and providing estimates for the model's parameters, the following equation-sets can be derived:

$$\frac{\partial(\ln L)}{\partial b_1} = \sum_{i=1}^{n} y_i - n b_1 - b_2 \sum_{i=1}^{n} x_i = 0 \tag{3.24}$$

$$\frac{\partial(\ln L)}{\partial b_2} = \sum_{i=1}^{n} x_i y_i - b_1 \sum_{i=1}^{n} x_i - b_2 \sum_{i=1}^{n} x_i^2 = 0 \tag{3.25}$$

$$\frac{\partial(\ln L)}{\partial \sigma_z} = n \cdot \sigma_z^2 - \sum_{i=1}^{n} (y_i - b_1 - b_2 x_i) = 0 \tag{3.26}$$

$$\frac{\partial(\ln L)}{\partial \tau_0} = \sum_{i=1}^{n} \frac{y_i - b_1 - b_2 x_i}{t_i - \tau_0} \tag{3.27}$$

From Eqs. 3.24–3.26, the solutions can be obtained as:

$$b_1 = \bar{y} - b_2 \bar{x} \tag{3.28}$$

$$b_2 = \frac{L_{xy}}{L_{xx}} \tag{3.29}$$

$$\sigma_z = \sqrt{\frac{Q}{n}} \tag{3.30}$$

where

$$\bar{x} = \frac{1}{n} \cdot \sum_{i=1}^{n} x_i \tag{3.31}$$

$$\bar{y} = \frac{1}{n} \sum_{i=1}^{n} y_i \tag{3.32}$$

$$Q = \sum_{i=1}^{n} (y_i - b_1 - b_2 x_i)^2 \tag{3.33}$$

$$L_{xx} = \sum_{i=1}^{n} x_i^2 - \frac{1}{n} \cdot \left(\sum_{i=1}^{n} x_i \right)^2 \tag{3.34}$$

$$L_{xy} = \sum_{i=1}^{n} x_i y_i - \frac{1}{n} \cdot \left(\sum_{i=1}^{n} x_i \right) \left(\sum_{i=1}^{n} y_i \right) \tag{3.35}$$

From Eqs. 3.19, 3.32, 3.33 and 3.35, it is clear that the constants b_1, b_2 and σ_z are binary functions of parameter τ_0. So it is necessary to first determine the solution of τ_0; the constants b_1, b_2 and σ_z can then be obtained. By solving Eq. 3.27 numerically, it is possible to have the solution of τ_0. The specific solution procedures are as follows:

(i) From Eq. 3.27, letting

$$H(\tau_0) = \sum_{i=1}^{n} \frac{y_i - b_1 - b_2 x_i}{t_i - \tau_0} \tag{3.36}$$

then it is possible to calculate the value of $H(\tau_0)$ for a given value of τ_0.

(ii) Based on Eq. 3.19, one has the value span of τ_0 as follows: $\tau_0 \in [0, t_{\min})$, where $t_{\min} = \min\{t_1, t_2, \ldots, t_n\}$ and t_i, $(i = 1, 2, \ldots, n)$ is the ith loading cycle in the experiment.

(iii) In case of a given set of initial prediction value of τ_0 (e.g., $\tau_0 = 0$) and a step length Δ, from Eqs. 3.28–3.30, the constants b_1, b_2 and σ_z can be determined and the value of function $H(\tau_0)$ can be then calculated by means of Eq. 3.36. With an iterative interpolation calculation, the τ_0–$H(\tau_0)$ relationship curve can be obtained, with τ_0 variable on the abscissa.

(iv) Using the relationship curve of τ_0–$H(\tau_0)$, it is possible to determine the point of intersection between the τ_0–$H(\tau_0)$ relation curve and the abscissa; this point is the solution of τ_0, where $H(\tau_0) = 0$.

(v) Again, from Eqs. 3.28–3.30, the predictions of constants b_1, b_2 and σ_z pertaining to the solution of τ_0 can be obtained.

From Eq. 2.66, the estimator $\hat{y}_p$ of safe logarithm crack growth life $y_p = \mu + u_p \sigma$ pertinent to reliability level p is

$$\hat{y}_p = \bar{y} + u_p\hat{k}\sigma_z = b_1 + b_2x + u_p\hat{k}\sigma_z \tag{3.37}$$

where u_p is the standard normal deviator corresponding to the reliability level of p.

Substituting Eq. 3.19 into Eq. 3.37 yields

$$\ln(t_p - \tau_0) = \ln C + m \cdot \ln a + u_p \cdot \hat{k}\sigma_z \tag{3.38}$$

Transformation of Eq. 3.38 gives

$$t_p = \tau_0 + Ca^m \cdot \exp\left[u_p\hat{k}\sigma_z\right] \tag{3.39}$$

Equation 3.39 is the crack propagation *a–t* curve determined by a randomized approach to the deterministic equation pertinent to a reliability level p. When $p = 50\%$, the probabilistic equation of crack propagation expressed by Eq. 3.39 degenerates to the deterministic equation (e.g., Eq. 3.5).

It is worth noting that in case of small sample numbers, there is an error between the predictions for sample percentile $\hat{y}_p$ of normal random variable Y, determined from Eq. 3.37, and the population true value. As a result, the confidence level of statistical results estimated from small samples needs to be analysed. The confidence level is a statistical variable correlated with sample size. Usually, the confidence levels are fixed and the amount of data determines the width of the corresponding confidence intervals. While the width of the corresponding confidence intervals is fixed and the amount of data determines the confidence levels, e.g., under a fixed width of corresponding confidence intervals, the fewer the samples, the less data items are available and thus the lower the confidence level in the statistical results. With larger sample numbers, there are more data items and thus the confidence level in the statistical results is higher. The sample percentile $\hat{y}_p$ of normal random variable Y pertinent to reliability level p, estimated using a large sample size, is near the population true value and the confidence levels are generally high. Thus it is not necessary for the confidence level of large sample sets to be analysed; only the reliability level need be estimated using Eq. 3.37.

From Eq. 2.95, the one-sided low limit of logarithmic fatigue life corresponding to the probability of survival of p and the confidence level of γ is

$$\hat{y}_{p\gamma} = \ln(\hat{t}_{p\gamma} - \tau_0) = (\bar{y} + u_p\hat{k}\sigma_z) - t_\gamma\sigma_z\sqrt{\frac{1}{n} + u_p^2(\hat{k}^2 - 1)} \tag{3.40}$$

Substituting Eq. 3.5 into Eq. 3.40 and transforming Eq. 3.40, it is possible to obtain

$$\hat{t}_{p\gamma} = \tau_0 + Ca^m \cdot \exp\left\{\sigma_z \cdot \left[\hat{k}u_p - t_\gamma\sqrt{\frac{1}{n} + u_p^2(\hat{k}^2 - 1)}\right]\right\} \tag{3.41}$$

Equation 3.41 is the formulation of the crack propagation *a–t* curve corresponding to a reliability level of p and a confidence level of γ. When $\gamma = 50\%$, then the probabilistic equation based on small sample numbers expressed by Eq. 3.41 can degenerate to Eq. 3.39. Additionally, while if $p = 50\%$ and $\gamma = 50\%$, then Eq. 3.41 can further degenerate to the deterministic equation (e.g., Eq. 3.5). Note that statistic y_p has an approximate normal distribution. This approximation is likely to be affected by the sparse data from the small samples that are usually available in fatigue analysis. Hence, the service life determined by Eq. 3.41 is approximate and not exact.

A randomized model of the deterministic equation can be also applied to reliability based S–N and $da/dN\text{–}\Delta K$ curves. Based on Eq. 1.7, the S–N curve of fatigue crack initiation can be described by a more general expression with three parameters as

$$N \cdot (S - S_0)^m = C \tag{3.42}$$

where S_0, m, C are undetermined material constants and N is the fatigue life.

The deterministic integral expression (3.42) of S–N curve for fatigue crack initiation is randomized and transformed into a natural logarithmic form as:

$$\ln N = \ln C - m \cdot \ln(S - S_0) + \ln W(S) \tag{3.43}$$

Again, letting $Y = \ln N$, $b_1 = \ln C$, $b_2 = -m$, $X = \ln(S - S_0)$, $Z(S) = \ln W(S)$, then Eq. 3.43 becomes

$$Y = b_1 + b_2 X + Z(S) \tag{3.44}$$

Using the analogy of Eqs. 3.39 and 3.41, it is possible to have

$$N_p \cdot (S - S_0)^m = C \cdot \exp\left[u_p \hat{k} \sigma_z\right] \tag{3.45}$$

$$N_{p\gamma} \cdot (S - S_0)^m = C \cdot \exp\left\{\sigma_z \cdot \left[\hat{k} u_p - t_\gamma \sqrt{\frac{1}{n} + u_p^2\left(\hat{k}^2 - 1\right)}\right]\right\} \tag{3.46}$$

Equation 3.46 is the probabilistic formulation of fatigue S–N curve corresponding to the reliability level of p and the confidence level of γ.

Now consider the fatigue crack growth law given in Eq. 1.40, re-written again below:

$$\frac{da}{dN} = C \cdot \Delta K^m \tag{3.47}$$

Using the analogy of Eq. 3.43, Eq. 3.47 can be transformed to yield:

$$\ln\left(\frac{da}{dN}\right) = \ln C + m \cdot \ln \Delta K + \ln W(\Delta K) \tag{3.48}$$

where $W(\Delta K)$ is the assumed stationary state-varying lognormal random process. Lin and Yang [23] assumed W as a time-varying random process. Yang and Chen [25] assumed $W(\Delta K)$ as lognormal random variable process. It is worth pointing out that the above assumptions need to be proved theoretically or validated by using experimental data.

Again, using the analogy of Eqs. 3.39 and 3.41, it is possible to obtain

$$\left(\frac{da}{dN}\right)_p = C\cdot(\Delta K)^m\cdot\exp\left[u_p\hat{k}\sigma_z\right] \tag{3.49}$$

$$\left(\frac{da}{dN}\right)_{p\gamma} = C\cdot(\Delta K)^m\cdot\exp\left\{\sigma_z\cdot\left[\hat{k}u_p + t_\gamma\sqrt{\frac{1}{n}+u_p^2(\hat{k}^2-1)}\right]\right\} \tag{3.50}$$

Example 3.1 In order to obtain the small sample number of fatigue crack growth *a–t* data sets, thirteen specimens made of LY12 aluminum alloy were used in fatigue tests subjected to 6906 load cycles in a block of the actual load spectra (shown in Fig. 3.1). The shapes and sizes of the specimens are shown in Fig. 3.2. The tests were carried out on an MTS-880–500KN fatigue test machine at a load frequency of 15 Hz, under the conditions of room temperature and atmospheric pressure. The variation of fatigue crack size with loading cycles was periodically observed and captured with a traveling optical microscope with a zoom lens. The initial crack size a_0 is 10.0 mm and the crack propagation size a was measured from the notch root (shown in Fig. 3.3). From Eqs. 3.28–3.36 and the dataset shown in Fig. 3.3, the parameters of the randomized *a–t* curve model can be estimated as

$$m = 0.2971, C = 4124.5, \tau_0 = 0.0, \sigma_z = 0.1876$$

From the annexed table of literature [26], it is possible to get

$$u_{0.99} = -2.326, t_{0.95}(\upsilon = 12) = 0.695, \hat{k}(n = 13) = 1.021$$

Thus, according to Eq. 3.41, the *a–t* curve formulae pertaining to reliability levels of 50% and 99% with a confidence level of 50% are respectively

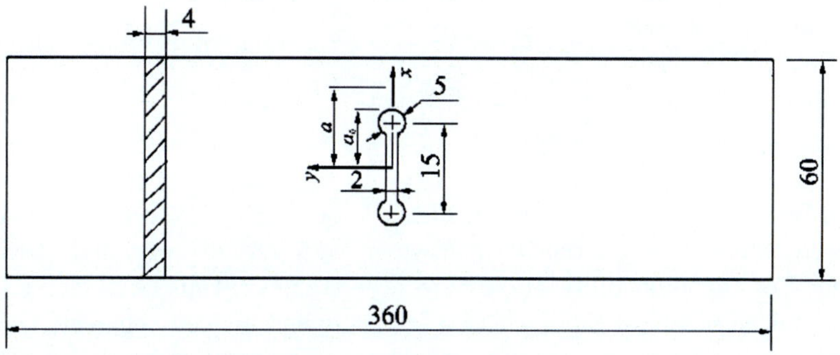

Fig. 3.1 Specimen dimension and shape (size unit:mm)

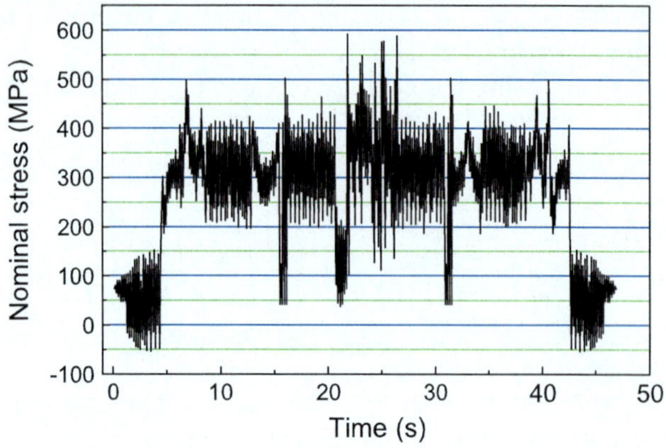

Fig. 3.2 Nominal stress spectrum

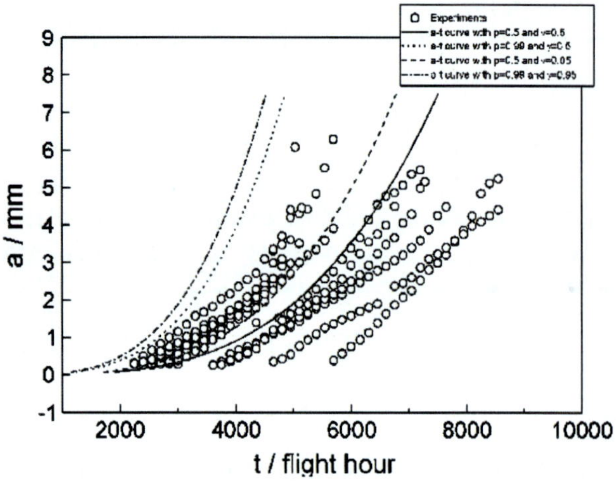

Fig. 3.3 Experiments and calculations of *a–t* curves with the reliability levels of 50%, 99% and the confidence levels of 50%, 95%

$$t = 4124.5 \times a^{0.2971} \tag{3.51}$$

$$t = 2725.5 \times a^{0.2971} \tag{3.52}$$

The *a–t* curve formulae pertaining to reliability levels of 50% and 99% with a confidence level of 95% are respectively

$$t = 2666.0 \times a^{0.2971} \tag{3.53}$$

$$t = 2484.0 \times a^{0.2971} \tag{3.54}$$

Equations 3.51–3.54 are shown in Fig. 3.3. From Fig. 3.3, it is seen that the calculated curves are reasonably consistent with the experimental data and the trends of the probability curves are also realistic. Thus, it is argued that the randomized model of Eq. 3.41 has adequately and logically characterized the physical characteristics and the phenomenological quantitative laws. Importantly, the parameters of this model can be estimated expediently and easily.

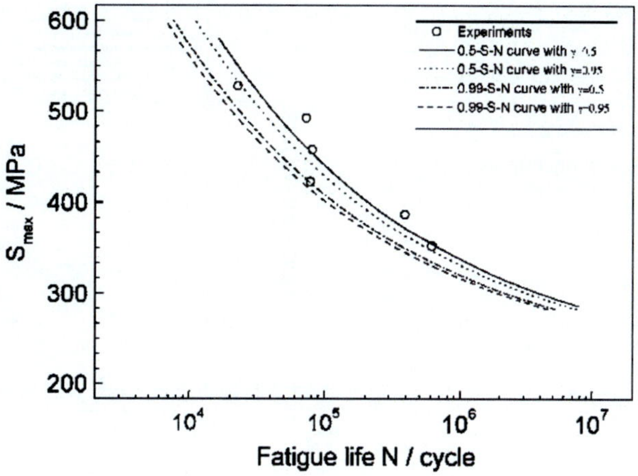

Fig. 3.4 Experiments and calculations of S–N curves with the reliability levels of 50%, 99% and the confidence levels of 50%, 95%

Example 3.2 This refers to a test programme for an S–N curve for an auricle junction at the root of a helicopter blade. The test sample consists of the auricle junction, bolts and the blade beam: the auricle junction and bolts are made of 40CrNiMoA alloyed steel while the blade beam is made of LD_2CS aluminium alloy. All tests were again carried out on MTS880−500KN servo-hydraulic machines. The experimental stress ratios was chosen to be $R = 0.1$. Each specimen was randomly sampled from the specimen sets to be subject to constant cyclic load in the axial direction at a loading frequency of 10 Hz under room temperature environment. For the fatigue tests, the stress levels were chosen as 352, 387, 423, 458, 493 and 528 MPa. One specimen was randomly sampled to be fatigue tested at each of the stress levels as shown in Fig. 3.4. From Eqs. 3.28–3.36 and the dataset shown in Fig. 3.4, it is possible to obtain the parameters of the randomized S–N curve model as

$$m = 3.72, \quad C = 5.39 \times 10^{13}, \quad S_0 = 217, \quad \sigma_z = 0.15$$

From the annexed table of literature [26], it is possible to get

$$u_{0.99} = -2.326, t_{0.95}(\upsilon = 5) = 0.727, \hat{k}(n = 6) = 1.051$$

Thus, according to Eq. 3.46, the S–N curve formulae pertaining to reliability levels of 50% and 99% with a confidence level of 50% are respectively

$$N = 5.39 \times 10^{13} \cdot (S - 217.0)^{-3.72} \tag{3.55}$$

$$N = 3.08 \times 10^{13} \cdot (S - 217.0)^{-3.72} \tag{3.56}$$

The S–N curve formulae pertaining to reliability levels of 50% and 99% with a confidence level of 95% are respectively

$$N = 4.49 \times 10^{13} \cdot (S - 217.0)^{-3.72} \tag{3.57}$$

$$N = 2.70 \times 10^{13} \cdot (S - 217.0)^{-3.72} \tag{3.58}$$

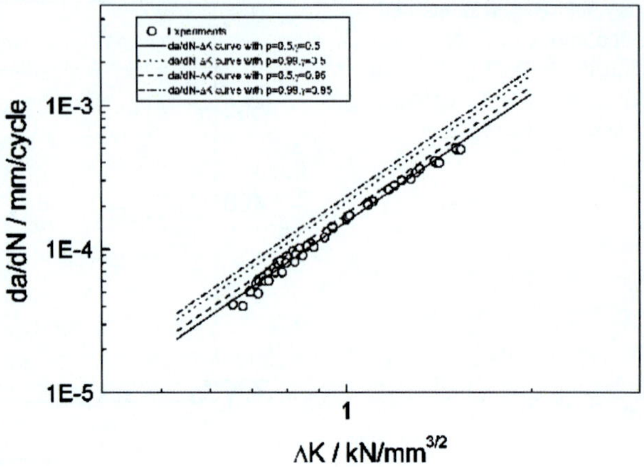

Fig. 3.5 Experiments and calculations of *da*/*dN*–Δ*K* curves with the reliability levels of 50%, 99% and the confidence levels of 50%, 95%

Equations 3.55–3.58 are shown in Fig. 3.4. It can be seen that the calculated curves are consistent with experimental data and the probability curves again are also realistic.

Example 3.3 Three CT specimens made of 40CrNiMo alloyed steel were used to determine fatigue crack growth rate. The thickness and width of the specimens are 15 mm and 80 mm respectively. The tests were carried out on an MTS-880–500KN fatigue test machine at a load frequency of 20 Hz, under the conditions of room temperature and atmospheric pressure. The experimental stress ratio was chosen to be $R = 0.1$ and the experimental dataset is shown in Fig. 3.5. From Eqs. 3.28–3.36 and the dataset shown in Fig. 3.5, the parameters of the randomized *da*/*dN*–Δ*K* curve model can be estimated as:

$$m = 2.9471, \quad C = 1.556 \times 10^{-4}, \quad \sigma_z = 0.1313$$

From the annexed table of literature [26], it is possible to get

$$u_{0.99} = -2.326, \quad t_{0.95}(\upsilon = 2) = 0.816, \quad \hat{k}(n = 3) = 1.086$$

Thus, according to Eq. 3.50, the *da*/*dN*–Δ*K* curve formulae pertaining to reliability levels of 50% and 99% with a confidence level of 50% are respectively

$$da/dN = 1.556 \times 10^{-4} \cdot (\Delta K)^{2.9471} \tag{3.59}$$

$$da/dN = 1.761 \times 10^{-4} \cdot (\Delta K)^{2.9471} \tag{3.60}$$

The *da*/*dN*–Δ*K* curve formulae pertaining to reliability levels of 50 and 99% with a confidence level of 95% are respectively

$$da/dN = 2.112 \times 10^{-4} \cdot (\Delta K)^{2.9471} \tag{3.61}$$

$$da/dN = 2.336 \times 10^{-4} \cdot (\Delta K)^{2.9471} \tag{3.62}$$

Equations 3.59–3.62 are shown in Fig. 3.5. In the same way, it can be seen that the calculated curves are consistent with experimental data. It is worth pointing out that the analysis results with higher confidence level are more conservative and safe from an engineering design viewpoint.

3.3 Single-Point Likelihood Method

Due to the stochastic nature of fatigue damage characterization, it helps to have large numbers of data sets to determine fatigue behaviour. A group testing method (GTM) shown in Fig. 3.6 together with the up-down method are recommended to determine a p–S–N curve in the SATM Standard E468-76 [27]; three groups of specimens corresponding to three different stress levels are fatigue tested. There are eight specimens in each group making a total of 24 for the GTM (shown in Fig. 3.2). Moreover, about 25 specimens need to be used for up-down testing. Thus about 50 specimens are required for the establishment of an S–N curve. In order to decrease the specimen numbers and test time, a single-point group testing method (SPGTM) [28–30] (shown in Fig. 3.7) was proposed for a p–S–N curve determination; seven specimens for the single-point tests and eight specimens for a group test are required to be fatigue tested. However, in some cases, due to cost and time constraints underpinning fatigue testing, only small numbers of samples can be provided. Hence, it is desirable to have a technique to address the paucity of data in assessing fatigue performance through a p–S–N curve. Based on Eq. 3.46, it is possible to obtain a single-point likelihood method (SPLM) [16] (shown in Fig. 3.8) for the assessment of structural fatigue performance from a minimal dataset and the specimen numbers for SPLM reduces to half that required for SPGTM. The SPLM was applied to experimental data, demonstrating the practical and effective use of the proposed model. It is worth pointing out that the randomized approach to the deterministic equation for a p–S–N curve was established in the above section based on the assumption of constant variance. In fact, it is proved from experience that, for the same specimens, fatigue life is a statistical variable correlated with the subjected stress level. If the stress level is higher, then the dispersion of fatigue life is less; conversely if the stress level is lower, then the

Fig. 3.6 GTM for S–N curve

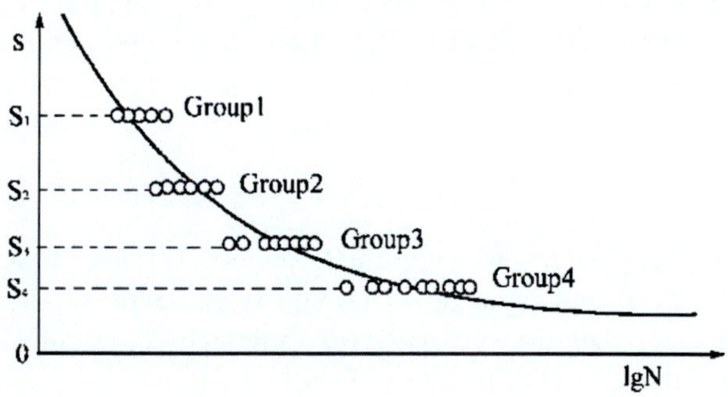

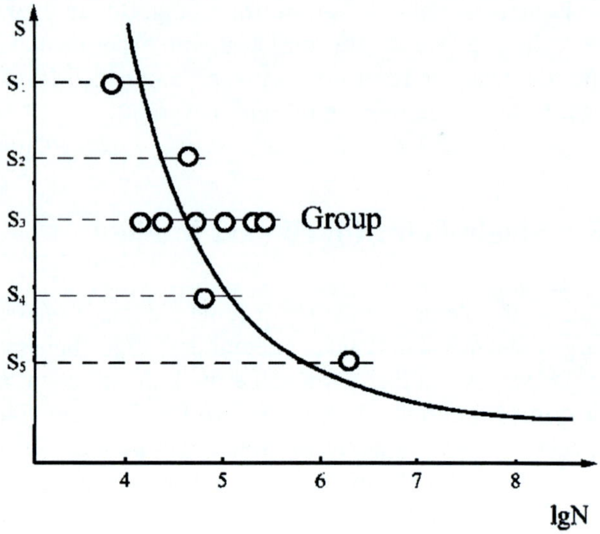

Fig. 3.7 SPGTM for S–N curve

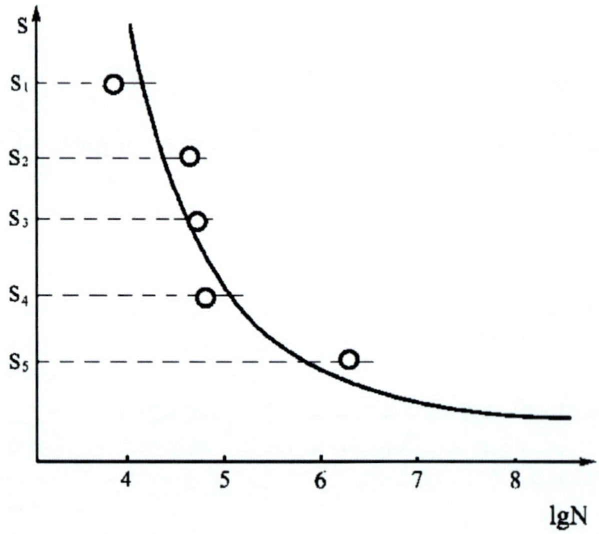

Fig. 3.8 SPLM for S–N curve

dispersion of fatigue life is greater. In order to take the heteroscedasticity of fatigue life into account, a new single-point likelihood method is presented for estimating p–S–N curve [17].

From Eq. 3.44, the lognormal model of S–N curve is

$$Y = b_1 + b_2 X + Z(S) \tag{3.63}$$

where $Z(S)$ is the lognormal random variable following a Gauss distribution $N[0, \sigma^2(S)]$ dependent on fatigue stress level S. The standard deviation $\sigma(S)$ to characterize the dispersion of fatigue life is generally assumed to be a linear relationship with ln S as [17, 30]:

$$\sigma(S) = c + d \cdot \ln S \tag{3.64}$$

Letting the standard deviation $\sigma(S)$ corresponding to fatigue stress $y_0 = \frac{1}{n}\sum_{i=1}^{n} \ln S_i$ be σ_0, Eq. 3.64 becomes

$$\sigma(S) = h \cdot \sigma_0 \tag{3.65}$$

where $h = 1 + g \cdot (\ln S - y_0)$, $g = d/\sigma_0$. From Eq. 3.63, the random variable Y follows Gauss distribution $N[b_1 + b_2x, \sigma^2(S)]$ and the natural logarithm of the above likelihood function can be written as

$$\ln L = -n \cdot \ln \sigma_0 - \frac{n}{2} \cdot \ln(2\pi) - \sum_{i=1}^{n} \ln(h_i) - \frac{1}{2\sigma_0^2} \cdot \sum_{i=1}^{n} \frac{(y_i - b_1 - b_2x_i)^2}{h_i^2} \tag{3.66}$$

According to the maximum likelihood principle and from Eq. 3.66, the following equations set can be derived:

$$\frac{\partial(\ln L)}{\partial b_1} = \sum_{i=1}^{n} \frac{y_i}{h_i^2} - b_1 \sum_{i=1}^{n} \frac{1}{h_i^2} - b_2 \sum_{i=1}^{n} \frac{x_i}{h_i^2} = 0 \tag{3.67}$$

$$\frac{\partial(\ln L)}{\partial b_2} = \sum_{i=1}^{n} \frac{x_i y_i}{h_i^2} - b_1 \sum_{i=1}^{n} \frac{x_i}{h_i^2} - b_2 \sum_{i=1}^{n} \frac{x_i^2}{h_i^2} = 0 \tag{3.68}$$

$$\frac{\partial(\ln L)}{\partial \sigma_0} = n \cdot \sigma_0^2 - \sum_{i=1}^{n} \frac{(y_i - b_1 - b_2x_i)^2}{h_i^2} = 0 \tag{3.69}$$

$$\frac{\partial(\ln L)}{\partial S_0} = \sum_{i=1}^{n} \frac{y_i - b_1 - b_2x_i}{(S - S_0)h_i^2} = 0 \tag{3.70}$$

$$\frac{\partial(\ln L)}{\partial g} = \sum_{i=1}^{n} \frac{\ln(S_i - y_0)}{h_i} \cdot \left[\frac{\sigma_0^2 h_i^2 - (y_i - b_1 - b_2x_i)^2}{\sigma_0^2 h_i^2}\right] = 0 \tag{3.71}$$

From Eqs. 3.67–3.69, the solutions can be obtained as

$$b_1 = \bar{y} - b_2\bar{x} \tag{3.72}$$

$$b_2 = \frac{L_{xy}}{L_{xx}} \tag{3.73}$$

$$\sigma_0 = \sqrt{\frac{Q}{n}} \tag{3.74}$$

where

$$\bar{x} = \frac{1}{m_0} \cdot \sum_{i=1}^{n} \frac{x_i}{h_i^2} \tag{3.75}$$

$$\bar{y} = \frac{1}{m_0}\sum_{i=1}^{n}\frac{y_i}{h_i^2} \tag{3.76}$$

$$Q = \sum_{i=1}^{n}\frac{(y_i - b_1 - b_2x_i)^2}{h_i^2} \tag{3.77}$$

$$m_0 = \sum_{i=1}^{n}\frac{1}{h_i^2} \tag{3.78}$$

$$L_{xx} = \sum_{i=1}^{n}\frac{x_i^2}{h_i^2} - \frac{1}{m_0}\cdot\left(\sum_{i=1}^{n}\frac{x_i}{h_i^2}\right)^2 \tag{3.79}$$

$$L_{xy} = \sum_{i=1}^{n}\frac{x_iy_i}{h_i^2} - \frac{1}{m_0}\cdot\left(\sum_{i=1}^{n}\frac{x_i}{h_i^2}\right)\left(\sum_{i=1}^{n}\frac{y_i}{h_i^2}\right) \tag{3.80}$$

From Eqs. 3.72–3.74, it is clear that the constants b_1, b_2 and σ_0 are binary functions of parameters g and S_0. So, by solving Eqs. 3.70 and 3.71 numerically, it is possible to obtain the solutions of g and S_0. And then the constants b_1, b_2 and σ_0 are determined.

Using the analogy of Eqs. 3.39 and 3.41, it can be shown that

$$N_p \cdot (S - S_0)^m \cdot S^{\left(-g \cdot u_p \hat{k} \sigma_0\right)} = C \cdot \exp\left[(1 - gy_0)u_p\hat{k}\sigma_0\right] \tag{3.81}$$

$$N_{p\gamma} \cdot (S - S_0)^m \cdot S^{\sigma_0 g\left[\hat{k}u_p - t_\gamma\sqrt{1/n + u_p^2(\hat{k}^2-1)}\right]} = C \cdot \exp\left\{\sigma_0(1 - g\cdot y_0)\left[\hat{k}u_p - t_\gamma\sqrt{1/n + u_p^2(\hat{k}^2 - 1)}\right]\right\} \tag{3.82}$$

Equation 3.82 is the probability formulation of fatigue S–N curve determined by heteroscedastic SPLM with a reliability level of p and a confidence level of γ.

The heteroscedastic SPLM established above is also capable of determining the generalized fatigue S–N surface. By analogy of Eq. 3.46 and using Eq. 1.18, the probability formulation of fatigue S–N surface determined by the heteroscedastic SPLM with a reliability level of p and a confidence level of γ can be obtained as:

$$N_{p\gamma} \cdot \left(\frac{\sigma_b}{\sigma_b - S_m}S_a - S_0\right)^m \cdot \left(\frac{\sigma_b}{\sigma_b - S_m}S_a\right)^{\sigma_0 g\left[\hat{k}u_p - t_\gamma\sqrt{1/n + u_p^2(\hat{k}^2-1)}\right]} = C \cdot \exp\left\{\sigma_0(1 - g\cdot y_0)\left[\hat{k}u_p - t_\gamma\sqrt{1/n + u_p^2(\hat{k}^2 - 1)}\right]\right\} \tag{3.83}$$

3.4 Generalized Constant Life Curve and Two-Dimensional Probability Distribution of Generalized Strength

As is well known, a precondition for fatigue and fracture reliability analysis and design is that the two-dimensional probability distributions of generalized strength (fatigue endurance limit, or fracture threshold) are known, while the precondition to induce the two-dimensional probability distribution of generalized strength is that a constant life curve is provided for establishing the relationship between binary variables of generalized strength. A new and phenomenological generalized fatigue constant life curve formula (1.5) for different materials is proposed in Sect. 1.4.

$$\frac{S_{\mathrm{a}}}{S_{-1}} + \left(\frac{S_{\mathrm{m}}}{\sigma_{\mathrm{b}}}\right)^{m} = 1 \tag{3.84}$$

where S_{a} is the fatigue endurance limit amplitude. S_{m} is fatigue endurance limit mean. σ_{b} is the fitted ultimate strength determined from experimental data. S_{-1} is the fitted fatigue endurance limit of the material under a symmetrical cyclic load from experimental data. m is a material constant obtained from experimental data. The fatigue endurance limit curve Eq. 3.84 is termed as the generalized fatigue constant life curve for describing the relationship between S_{a} and S_{m}. On the basis of fatigue endurance limit data ($S_{\mathrm{a}i}$, $S_{\mathrm{m}i}$) determined by the up-and-down method, an $S_{\mathrm{a}}-S_{\mathrm{m}}$ curve can be determined.

Using the analogy of Eq. 3.84, a phenomenological generalized fracture constant life curve formula for all kinds of materials can be written to express the relationship between fracture threshold range ΔK_{th} and mean $K_{m\mathrm{th}}$ as:

$$\frac{\Delta K_{\mathrm{th}}}{\Delta K_{-1}} + \left(\frac{K_{m\,\mathrm{th}}}{K_{1\mathrm{C}}}\right)^{m} = 1 \tag{3.85}$$

where ΔK_{th} is the fracture threshold range. ΔK_{-1} is the fitted fracture threshold under symmetrical cyclic loading from experimental data. $K_{1\mathrm{C}}$ is the fitted fracture toughness of material from experimental data. $K_{m\mathrm{th}}$ is the fracture threshold mean.

Equation 3.85 is termed as the generalized fracture constant life curve. ΔK_{-1}, $K_{1\mathrm{C}}$ and m are gained from fracture experiments. Based on fracture threshold data (ΔK_{th}, $K_{m\mathrm{th}}$), a fracture constant life $\Delta K_{\mathrm{th}}-K_{m\mathrm{th}}$ curve can be fitted. Equations 3.84 and 3.85 are power functions with three parameters and can be determined by using the analogy of Eq. 3.41

Transforming Eq. 3.84 gives

$$\left(\frac{S_{\mathrm{m}}}{\sigma_{\mathrm{b}}}\right)^{m} = \frac{S_{-1} - S_{\mathrm{a}}}{S_{-1}} \tag{3.86}$$

Equation 3.86 is randomized and transformed into a natural logarithmic form as

$$\ln S_{\mathrm{m}} = \ln \sigma_{\mathrm{b}} - \frac{1}{m}\ln S_{-1} + \frac{1}{m}\ln(S_{-1} - S_{\mathrm{a}}) + \ln W(S_{\mathrm{a}}) \tag{3.87}$$

where the random factor, $W(S_{\mathrm{a}})$ is a non-negative, stationary lognormal random process of space with a median value of 1.0 and a standard deviation σ_w.

Letting $Y = \ln S_{\mathrm{m}}$, $b_1 = \ln \sigma_{\mathrm{b}} - \frac{1}{m}\ln S_{-1}$, $b_2 = \frac{1}{m}$, $X = \ln\,[S_{-1} - S_{\mathrm{a}}]$, $Z(S_{\mathrm{a}}) = \ln W(S_{\mathrm{a}})$, then Eq. 3.87 becomes

$$Y = b_1 + b_2 X + Z(S_{\mathrm{a}}) \tag{3.88}$$

where $Z(S_{\mathrm{a}})$ is the state-varying Gauss random process with a median value of 1.0 and a standard deviation σ_z. From Eq. 3.88, stochastic process Y is Gauss random variable with a median value of $b_1 + b_2 x$ and a standard deviation σ_z. Thus likelihood function of random variable Y is obtained and maximum likelihood equation set can be derived as

$$\sum_{i=1}^{n} y_i - nb_1 - b_2 \sum_{i=1}^{n} x_i = 0 \tag{3.89}$$

$$\sum_{i=1}^{n} x_i y_i - b_1 \sum_{i=1}^{n} x_i - b_2 \sum_{i=1}^{n} x_i^2 = 0 \tag{3.90}$$

$$n \cdot \sigma_z^2 - \sum_{i=1}^{n} (y_i - b_1 - b_2 x_i)^2 = 0 \tag{3.91}$$

$$\sum_{i=1}^{n} \frac{y_i - b_1 - b_2 x_i}{S_{-1} - S_{\mathrm{a}}} = 0 \tag{3.92}$$

Solving Eqs. 3.89–3.91 yield

$$b_1 = \bar{y} - b_2 \bar{x} \tag{3.93}$$

$$b_2 = \frac{L_{xy}}{L_{xx}} \tag{3.94}$$

$$\sigma_z = \sqrt{\frac{Q}{n}} \tag{3.95}$$

where

$$\bar{x} = \frac{1}{n} \cdot \sum_{i=1}^{n} x_i$$

$$\bar{y} = \frac{1}{n} \sum_{i=1}^{n} y_i$$

$$Q = \sum_{i=1}^{n} (y_i - b_1 - b_2 x_i)^2$$

$$L_{xx} = \sum_{i=1}^{n} x_i^2 - \frac{1}{n} \cdot \left(\sum_{i=1}^{n} x_i \right)^2$$

$$L_{xy} = \sum_{i=1}^{n} x_i y_i - \frac{1}{n} \cdot \left(\sum_{i=1}^{n} x_i \right) \left(\sum_{i=1}^{n} y_i \right)$$

From the above equations, it can be shown that the undetermined parameters b_1, b_2 and σ_z are associated with the unknown variable S_{-1}. So, adopting the same method for solving Eq. 3.27 numerically and using Eq. 3.92, it is possible to have the solution of unknown variable S_{-1} and to then obtain the solutions of b_1, b_2 and σ_z.

Using the analogy of Eqs. 3.39 and 3.41, it can be shown that

$$(S_{\mathrm{m}})_p = \frac{\sigma_{\mathrm{b}}}{S_{-1}^{\frac{1}{m}}} \cdot (S_{-1} - S_{\mathrm{a}})^{\frac{1}{m}} \cdot \exp\left(u_p \cdot \hat{k} \sigma_z\right) \tag{3.96}$$

$$(S_{\mathrm{m}})_{p\gamma} = \frac{\sigma_b}{S_{-1}^{\frac{1}{m}}} \cdot (S_{-1} - S_{\mathrm{a}})^{\frac{1}{m}} \cdot \exp\left\{ \sigma_z \cdot \left[\hat{k} u_p - t_\gamma \sqrt{\frac{1}{n} + u_p^2 (\hat{k}^2 - 1)} \right] \right\} \tag{3.97}$$

Equation 3.97 is termed as the generalized fatigue constant life curve with a reliability level of p and a confidence level of γ. Similarly, it is possible to generate the generalized fracture constant life curve with a reliability level of p and a confidence level of γ.

$$(K_{m\mathrm{th}})_{p\gamma} = \frac{K_{1C}}{\Delta K_{-1}^{\frac{1}{m}}} \cdot (\Delta K_{-1} - \Delta K_{\mathrm{th}})^{\frac{1}{m}} \cdot \exp\left\{ \sigma_z \cdot \left[\hat{k} u_p - t_\gamma \sqrt{\frac{1}{n} + u_p^2 (\hat{k}^2 - 1)} \right] \right\} \tag{3.98}$$

Equations 3.97 and 3.98 are generalized fatigue and fracture constant life curves respectively. Based on the definition of conditional probability, it is possible to prove that at a constant life curve, the probability of generalized strength amplitude (or range) at a specific generalized strength mean equals that of generalized strength mean at a specific generalized strength amplitude (or range). Therefore, from Eqs. 3.97 and 3.98, the probability distributions of generalized strength amplitude (or range) at a specific generalized strength mean and generalized strength mean at a specific generalized strength amplitude (or range) are obtained respectively.

As is well known, the S–N (or stress-life) curve implies that fatigue failure in the material occurs after N cycles of loading at stress level S, i.e. the fatigue strength of material subjected N cycles of cyclic loading is stress level S. From Eq. 1.5, it is shown that under a specific stress ratio R, S–N curve can be expressed as

$$S_{\mathrm{m}} \cdot N^{m_1} = C \tag{3.99}$$

The logarithm form of Eq. 3.99 becomes

$$Y = a_0 + b_0 X \tag{3.100}$$

where

$$Y = \ln S_{\mathrm{m}}, \quad X = \ln N, \quad a_0 = \frac{1}{m} \ln C, \quad b_0 = -\frac{1}{m} \tag{3.101}$$

Long-term experience shows that the logarithmic fatigue life follows normal distribution [31–33], or $X \sim N(\mu_x, \sigma_x^2)$, thus, from Eq. 3.100, random variable $Y \sim N(\mu_y, \sigma_y^2)$, here

$$\mu_y = a_0 + b_0 \mu_x \tag{3.102}$$

$$\sigma_y = b_0 \sigma_x \tag{3.103}$$

Similarly, the logarithm expression of Eq. 3.84 can be written as

$$Z = a + bY \tag{3.104}$$

where

$$Y = \ln S_{\mathrm{m}}, Z = \ln(S_{-1} - S_{\mathrm{a}}), a = \ln S_{-1} - m \ln \sigma_{\mathrm{b}}, b = m \tag{3.105}$$

From Eq. 3.104, it can be shown that random variable $Z \sim N(\mu_z, \sigma_z^2)$, has

$$\mu_z = a + b\mu_y \tag{3.106}$$

$$\sigma_z = b\sigma_y \tag{3.107}$$

According to the definition of two-dimensional joint probability density function of random variables, the two-dimensional joint probability density function of random variables Y and Z can be deduced as

$$f(y,z)=\frac{1}{2\pi\sigma_y\sigma_z\sqrt{1-r^2}}\exp\left\{\frac{-1}{2(1-r^2)}\left[\left(\frac{y-\mu_y}{\sigma_y}\right)^2-\frac{2r(y-\mu_y)(z-\mu_z)}{\sigma_y\sigma_z}+\left(\frac{z-\mu_z}{\sigma_z}\right)^2\right]\right\} \tag{3.108}$$

where r is the linearly dependent coefficient determined from following equation as

$$r=\frac{L_{YZ}}{\sqrt{L_{YY}L_{ZZ}}}=\frac{\sum_{i=1}^{n}y_iz_i-\frac{1}{n}\left(\sum_{i=1}^{n}y_i\right)\left(\sum_{i=1}^{n}z_i\right)}{\sqrt{\left[\sum_{i=1}^{n}y_i^2-\frac{1}{n}\left(\sum_{i=1}^{n}y_i\right)^2\right]\left[\sum_{i=1}^{n}z_i^2-\frac{1}{n}\left(\sum_{i=1}^{n}z_i\right)^2\right]}} \tag{3.109}$$

From the definitions (3.101) and (3.105) of random variables Y and Z, it is possible to have

$$S_a=S_{-1}-\exp(Z) \tag{3.110}$$

$$S_m=\exp(Y) \tag{3.111}$$

Equations 3.110 and 3.111 must have continuous partial derivatives. From Eqs. 3.110 and 3.111, one has the Jacobi determinant function for the transformation as

$$J=\begin{vmatrix}\frac{\partial y}{\partial S_a} & \frac{\partial y}{\partial S_m}\\ \frac{\partial z}{\partial S_a} & \frac{\partial z}{\partial S_m}\end{vmatrix}=\begin{vmatrix}0 & \frac{1}{S_m}\\ \frac{-1}{S_{-1}-S_a} & 0\end{vmatrix}=\frac{1}{S_m(S_{-1}-S_a)} \tag{3.112}$$

According to the definition of probability density function of random variable function, from Eqs. 3.108 and 3.112, the two-dimensional joint probability density function of fatigue strength (S_a, S_m) can be deduced as

$$\begin{aligned}g(S_a,S_m)&=f[\ln S_m,\ln(S_{-1}-S_a)]\cdot|J|\\&=\frac{1}{2\pi\sigma_y\sigma_zS_m(S_{-1}-S_a)\sqrt{1-r^2}}\\&\quad\times\exp\left\{\frac{-1}{2(1-r^2)}\left[\left(\frac{\ln S_m-\mu_y}{\sigma_y}\right)^2-\frac{2r(\ln S_m-\mu_y)(\ln(S_{-1}-S_a)-\mu_z)}{\sigma_y\sigma_z}\right.\right.\\&\quad\left.\left.+\left(\frac{\ln(S_{-1}-S_a)-\mu_z}{\sigma_z}\right)^2\right]\right\}\end{aligned} \tag{3.113}$$

or

$$g(S_a, S_m) = \frac{1}{2\pi b_0^2 b\sigma_x^2 S_m (S_{-1} - S_a)\sqrt{1 - r^2}} \times \exp\left\{\frac{-1}{2(1 - r^2)}\left[\left(\frac{\ln S_m - a_0 - b_0\mu_x}{b_0\sigma_x}\right)^2 - \frac{2r(\ln S_m - a_0 - b_0\mu_x)(\ln(S_{-1} - S_a) - a - a_0 b - b_0 b\mu_x)}{b_0^2 b\sigma_x^2} + \left(\frac{\ln(S_{-1} - S_a) - a - a_0 b - b_0 b\mu_x}{b_0 b\sigma_x}\right)^2\right]\right\} \quad (3.114)$$

Using the analogy of Eq. 3.114, it can be shown that

$$g(\Delta K_{th}, K_{mth}) = \frac{1}{2\pi b_0^2 b\sigma_x^2 K_{mth}(\Delta K_{-1} - \Delta K_{th})\sqrt{1 - r^2}} \times \exp\left\{\frac{-1}{2(1 - r^2)}\left[\left(\frac{\ln K_{mth} - a_0 - b_0\mu_x}{b_0\sigma_x}\right)^2 - \frac{2r(\ln K_{mth} - a_0 - b_0\mu_x)(\ln(\Delta K_{-1} - \Delta K_{th}) - a - a_0 b - b_0 b\mu_x)}{b_0^2 b\sigma_x^2} + \left(\frac{\ln(\Delta K_{-1} - \Delta K_{th}) - a - a_0 b - b_0 b\mu_x}{b_0 b\sigma_x}\right)^2\right]\right\} \quad (3.115)$$

Equation 3.115 is the two-dimensional joint probability density function of fracture strength (ΔK_{th}, K_{mth}).

In order to validate the generalized constant life fatigue curve and the two-dimensional probability distribution of generalized strength, fatigue limit tests were conducted using the two-dimensional up-and-down method and two-dimensional fracture threshold tests on LY11CS aluminium alloy and 40CrNiMoA alloyed steel specimens. These are outlined in the two examples that follow.

Example 3.4 Test specimens elaborated to simulate the helicopter's rotor wings made of LY11CS aluminium alloy (here CS represents the state and heating processing of solution treatment and artificial aging of A-alloy) with a thickness of 5 mm and a stress concentration factor of $K = 1$, were subjected to loading at a frequency of 130 Hz. Three sets of up-and-down tests (shown in Figs. 3.9, 3.10 and 3.11) pertaining to three different mean stress levels of 58, 69 and 82 MPa respectively at the given life of $N = 10^7$ were carried out on a high frequency fatigue machine of AMSKR1478-10 at room temperature. As shown in Figs. 3.9, 3.10 and 3.11, 21, 25 and 20 specimens were used in up-and-down tests at three different mean stress levels of 58, 69 and 82 MPa respectively. The experimental results are listed in Table 3.1.

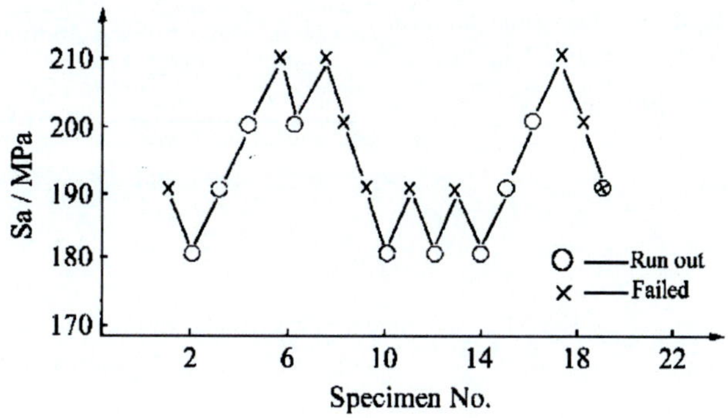

Fig. 3.9 Up-and-down Method Test at $N = 10^7$ and $S_m = 58$ MPa for LY11CS A-alloy

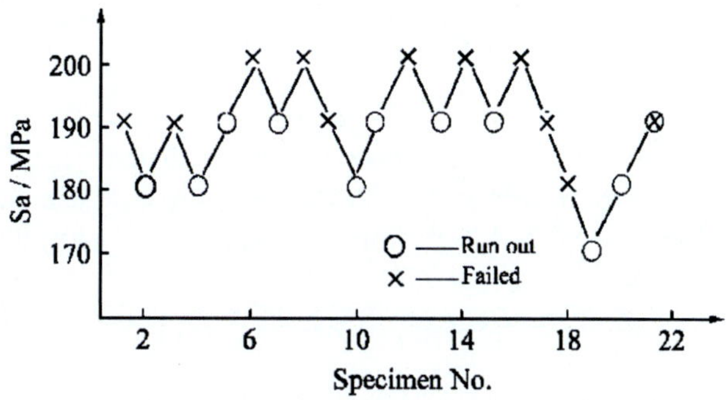

Fig. 3.10 Up-and-down Method Test at $N = 10^7$ and $S_m = 69$ MPa for LY11CS A-alloy

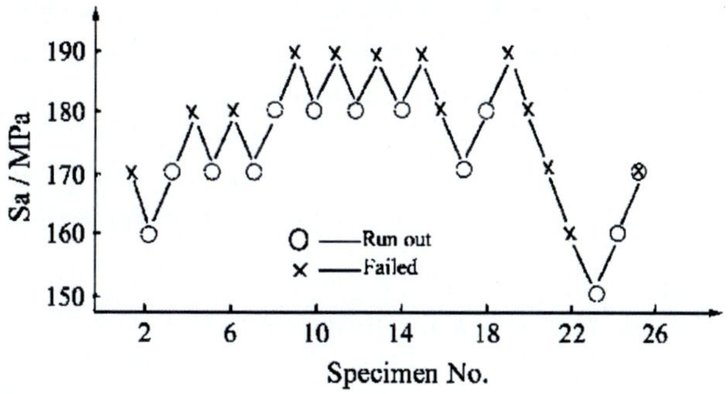

Fig. 3.11 Up-and-down Method Test at $N = 10^7$ and $S_m = 82$ MPa for LY11CS A-alloy

From Eqs. 3.92 and 3.95 and using the experimental data listed in Table 3.1, the median generalized fatigue constant life S_a–S_m curve equation can be determined.

$$\frac{S_a}{197.8} + \left[\frac{S_m}{126.5}\right]^{5.07} = 1 \tag{3.116}$$

And from Eq. 3.109, the linear correlation coefficient r is 0.9942. By using the above parameters and Eq. 3.105, one has

$$a = -19.25$$
$$b = 5.07$$

Table 3.1 Experimental data at $N = 10^7$ for LY11CS

Mean stress (S_m/MPa)	Stress amplitude (S_{ai}/MPa)	Sample size	Mean of stress amplitude (S_a/MPa)
58	205	3	194.0
	195	3	
	185	4	
69	195	5	188.64
	185	5	
	175	1	
82	185	5	175.83
	175	4	
	165	2	
	155	1	

From the experimental results listed in Table 3.1 and parameter $S_{-1} = 197.8$ MPa in Eq. 3.116, the mean and standard variation of random variable $Z = \ln\,(S_{-1} - S_a)$ at a specific mean strength of $S_m = 82$ MPa can be obtained.

$$\mu_z = 3.0$$
$$\sigma_z = 0.4403$$

By means of Eqs. 3.106 and 3.107, it is possible to determine the mean and standard variation of random variable $Y = \ln S_m$ at a specific fatigue strength amplitude of $S_a = 175.83$ MPa as

$$\mu_y = 4.4$$
$$\sigma_y = 0.087$$

Substituting the above parameters into Eq. 3.114 gives the two-dimensional joint probability density function of (S_a, S_m) as

$$\begin{aligned} g(S_a, S_m) = {} & \frac{38.6}{S_m(197.8 - S_a)} \times \exp\left\{-43.2 \times \left[\left(\frac{\ln S_m - 4.4}{0.087}\right)^2\right.\right. \\ & - 51.9 \times (\ln S_m - 4.4)(\ln(197.8 - S_a) - 3.0) \\ & \left.\left. + \left(\frac{\ln(197.8 - S_a) - 3.0}{0.4403}\right)^2\right]\right\} \end{aligned} \quad (3.117)$$

Equation 3.117 is shown in Figs. 3.12, 3.13 and 3.14. From Fig. 3.12, it is clear that a 'hump' in the two-dimensional probability density surface exists. The projection of this profile of the two-dimensional probability density surface becomes the constant life curve on the S_a–S_m plane (shown in Fig. 3.13). Figure 3.14 indicates that the cross-sectional views of the two-dimensional probability density surface are the probability density curves of logarithmic normal

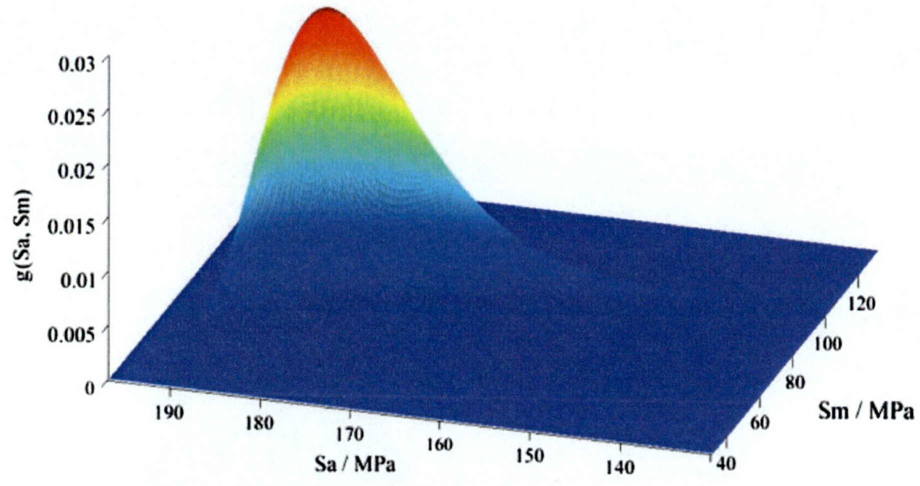

Fig. 3.12 Two-dimensional joint distribution density surface of fatigue endurance limit at $N = 10^7$ for LY11CS A-alloy

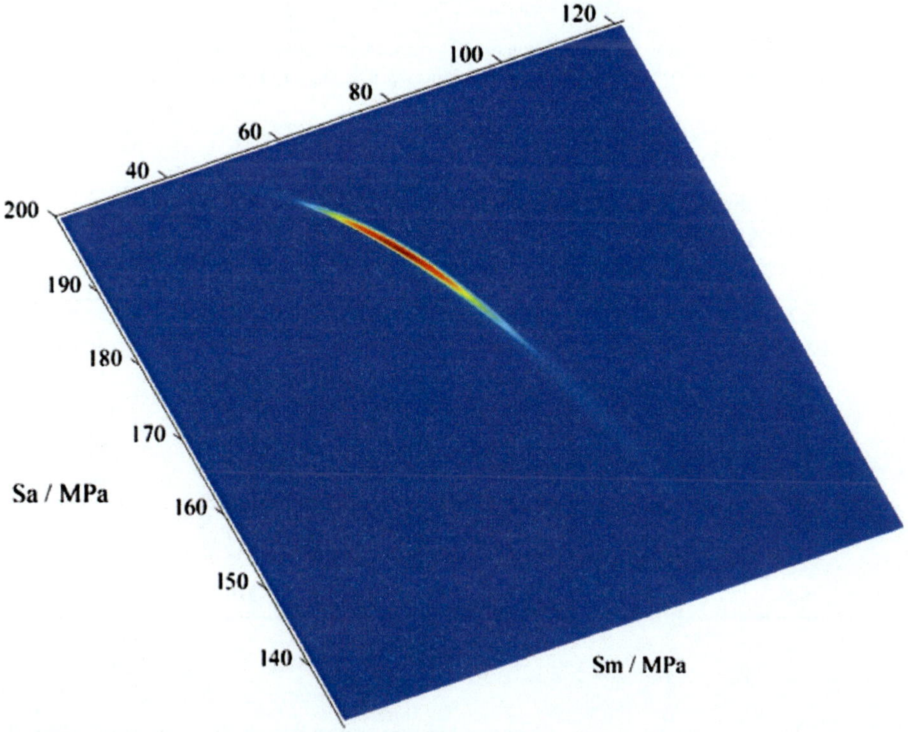

Fig. 3.13 Projecting figure of two-dimensional joint distribution density surface of fatigue endurance limit at $N = 10^7$ on the plane of coordinates S_a and S_m for LY11CS A-alloy

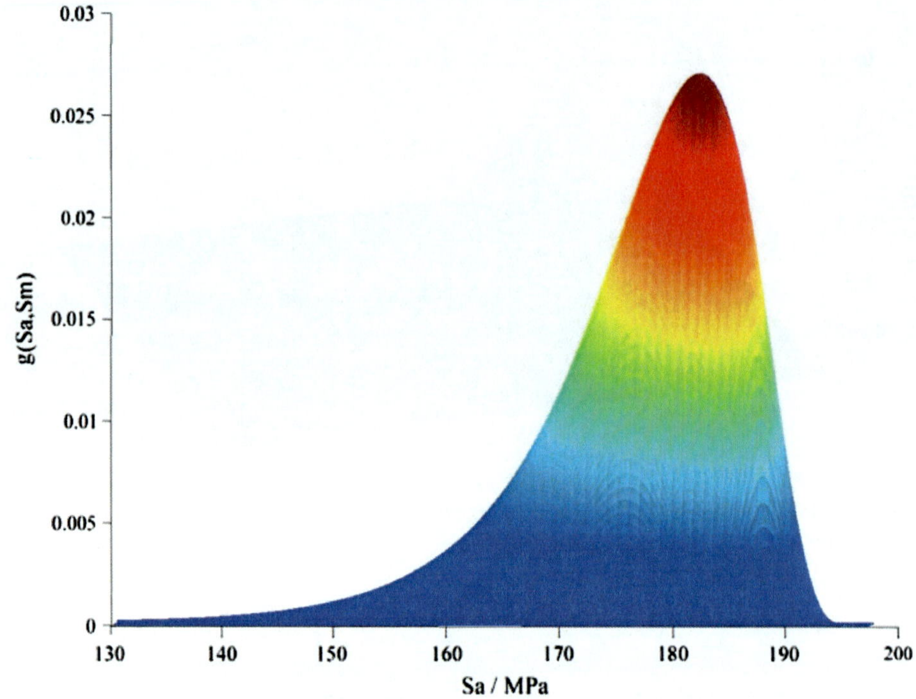

Fig. 3.14 Projecting figure of two-dimensional joint distribution density surface of fatigue endurance limit at $N = 10^7$ on the plane of coordinates $f(S_a, S_m)$ and S_a for LY11CS A-alloy

Table 3.2 Experimental data of two-dimensional fracture threshold ΔK_{th} (unit: MPa mm$^{1/2}$)

	LY11CS aluminium alloy			40CrNiMoA alloyed-steel		
No.	$K_{mth} = 210$	$K_{mth} = 247$	$K_{mth} = 306$	$K_{mth} = 584$	$K_{mth} = 707$	$K_{mth} = 920$
1	66.29	60.72	53.33	181.42	164.23	145.95
2	68.71	61.25	53.34	188.04	166.21	145.98
3	71.46	61.88	55.68	195.57	168.13	152.38
4	74.83	66.31	57.89	204.79	181.17	158.43
5		66.92	61.07		183.94	167.13
6		63.21			182.36	
7		66.13			184.21	
8		70.64			189.34	
9		71.09			192.77	
Mean	70.32	65.46	56.26	192.46	179.15	153.98

distribution on the plane of coordinates $g(S_a, S_m)$ and S_a or S_m. It is thus argued that the two-dimensional joint probability density function formulation has adequately and logically characterized the physical characteristics.

Table 3.3 Calculated parameter results of two-dimensional fracture threshold

Parameter	LY11CS	40CrNiMoA	Parameter	LY11CS	40CrNiMoA
m	1.69	1.40	b	1.69	1.4
ΔK_{-1}	86.12	235.71	μ_z	3.2	4.02
K_{1C}	573.53	1960.12	σ_z	0.1952	0.1815
r	0.9976	0.9945	μ_y	5.6	6.55
a	−6.28	−5.15	σ_y	0.1155	0.1296

Example 3.5 Compact tension (CT) specimens with a thickness of 25 mm and a width of 50 mm were made from two kinds of material, LY11CS aluminium alloy and 40CrNiMoA alloyed-steel respectively. Two-dimensional fracture threshold tests were carried out on MTS880−50KN servo-hydraulic machines where frequency and stress ratio could be changed and the load or displacement could be controlled by a load cell, at room temperature and moisture. The loading frequency range was chosen to be 15 Hz. The experimental data of LY11CS aluminium alloy and 40CrNiMoA alloyed-steel are shown in Table 3.2.

Using the same method as in the Example 3.4 and using the fracture threshold data listed in Table 3.2, the parameters of generalized fracture constant life curve equation and two-dimensional joint probability density function of (ΔK_{th}, $K_{m\text{th}}$) can be determined respectively (shown in Table 3.3).

Substituting the parameters listed in Table 3.3 into Eq. 3.115 gives the two-dimensional joint probability density functions of (ΔK_{th}, $K_{m\text{th}}$) of LY11CS aluminium alloy and 40CrNiMoA alloyed-steel respectively

$$g(\Delta K_{\text{th}}, K_{m\text{th}}) = \frac{101.95}{K_{m\text{th}}(86.12 - \Delta K_{\text{th}})} \times \exp\left\{-104.29 \times \left[\left(\frac{\ln K_{m\text{th}} - 5.6}{0.1155}\right)^2 - 88.50 \times (\ln K_{m\text{th}} - 5.6)(\ln(86.12 - \Delta K_{\text{th}}) - 3.2) + \left(\frac{\ln(86.12 - \Delta K_{\text{th}}) - 3.2}{0.1952}\right)^2\right]\right\} \tag{3.118}$$

$$g(\Delta K_{\text{th}}, K_{m\text{th}}) = \frac{64.60}{K_{m\text{th}}(235.71 - \Delta K_{\text{th}})} \times \exp\left\{-45.58 \times \left[\left(\frac{\ln K_{m\text{th}} - 6.55}{0.1296}\right)^2 - 84.57 \times (\ln K_{m\text{th}} - 6.55)(\ln(235.71 - \Delta K_{\text{th}}) - 4.02) + \left(\frac{\ln(235.71 - \Delta K_{\text{th}}) - 4.02}{0.1815}\right)^2\right]\right\} \tag{3.119}$$

Equation 3.119 is shown in Fig. 3.15. Figure 3.15 shows that the two-dimensional joint probability density function formulation has adequately and logically characterized the physical characteristics with reasonable accuracy.

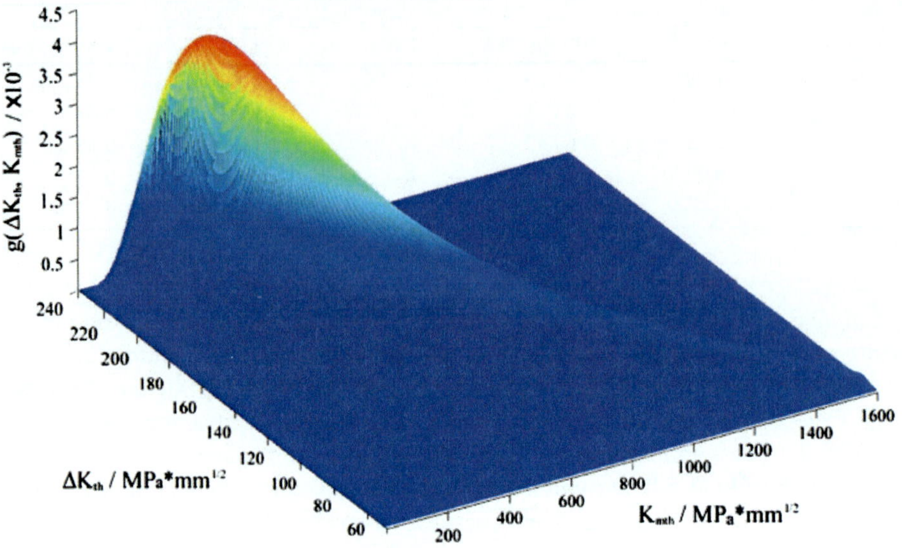

Fig. 3.15 Two-dimensional joint distribution density surface of fracture threshold value for 40CrNiMoA alloyed-steel

3.5 Full-range S–N Curve and Crack Growth Rate Curve with Four Parameters

From Eq. 1.7, it is obvious that the three-parameter power function formula is suitable for mid-life and long-life ranges of S–N curve but is inapplicable for the low-life range. Though the S–N curve Eq. 1.9 covers the overall life range, it is difficult to estimate its parameters. According to Eq. 1.9, a power formula of S–N curve formula covering the overall life range with four parameters is developed as

$$\left(\frac{S_{\mathrm{u}} - S_0}{S - S_0}\right) = 10^{C(\log N)^m} \tag{3.120}$$

where S_0, m, C and S_{u} are the material constants. S_u is the ultimate strength, S_0 is the fitting fatigue limit or reference fatigue strength and m is exponent material related constant. Equation 3.120 has the physical properties as follows:

If $N = 1$, then $S = S_u$

If $N = \infty$, then $S = S_0$

Greater the fatigue stress S, shorter the fatigue life N.

In order to estimate the parameters of Eq. 3.120, taking the logarithmic transformation of Eq. 3.120 twice yields

$$Y = b_1 + b_2 X \tag{3.121}$$

where $Y = \ln[\log(S_u - S_0) - \log(S - S_0)]$, $b_1 = \ln C$, $b_2 = m$ and $X = \ln(\log N)$.

On the basis of linear regression of mathematical statistics, from Eq. 3.121, it can be shown that

$$b_1 = \bar{y} - b_2\bar{x} \tag{3.122}$$

$$b_2 = \frac{L_{xy}}{L_{xx}} \tag{3.123}$$

$$r(S_u, S_0) = \frac{L_{xy}}{\sqrt{L_{xx}L_{yy}}} \tag{3.124}$$

where

$$\bar{x} = \frac{1}{n} \cdot \sum_{i=1}^{n} x_i$$

$$\bar{y} = \frac{1}{n} \sum_{i=1}^{n} y_i$$

$$L_{xx} = \sum_{i=1}^{n} x_i^2 - \frac{1}{n} \cdot \left(\sum_{i=1}^{n} x_i \right)^2$$

$$L_{xy} = \sum_{i=1}^{n} x_i y_i - \frac{1}{n} \cdot \left(\sum_{i=1}^{n} x_i \right) \left(\sum_{i=1}^{n} y_i \right)$$

Equations 3.122 and 3.123 show that b_1 and b_2 are binary functions of undetermined parameters S_u and S_0. As a result, by means of the optimization method of linear dependent coefficient $r(R_0, S_0)$ (or maximum absolute value of linear dependent coefficient $r(R_0, S_0)$), it is possible to have the solutions of S_u and S_0, and then to determine the values of constants b_1 and b_2 from Eqs. 3.122 and 3.123. S_u and S_0 are solved from formula (3.124) as follows in a four-stage process.

(a) Firstly, the value ranges of S_u and S_0 should be predicted as

$$S_u \in (S_{u\max}, \sigma_{0.2}], S_0 \in [0, S_{0\min})$$

where $S_{0\min} = \min\ \{S_1, S_2, \ldots, S_n\}$, $S_{u\max} = \max\ \{S_1, S_2, \ldots, S_n\}$ and S_i, $(i = 1, 2, \ldots, n)$ is the fatigue stress from test data. $\sigma_{0.2}$ is the material yield limit.

(b) Secondly, giving a set of initial prediction values and the step lengths Δ_1 and Δ_2 of S_u and S_0, by using Eq. 3.124, the values of S_u and S_0 corresponding to the maximum absolute value of linear dependent coefficient $r(R_0, S_0)$ can then be searched and obtained.

(c) Thirdly, based on the values of S_u and S_0 determined in the above calculation programme and using Eqs. 3.122 and 3.123, one has the solutions of b_1 and b_2.

As is well known, the ultimate strength S_u is a random variable. From Eq. 3.120, it can be proved that fatigue strength S at a given fatigue life N is also a random variable, and the probabilities of ultimate strength S_u and fatigue strength S are same. Fatigue strength S_p corresponding to reliability level p is

$$S_p = 10^{-C(\log N)^m}\left(S_{up} - S_0\right) + S_0 \tag{3.125}$$

Equation 3.125 is the probability formula of four-parameter S–N curve covering the overall life range. In general, ultimate strength S_u follows a two-parameter Weibull distribution [34], alternatively,

$$F_{S_\mathrm{u}}(x) = P[S_\mathrm{u} \le x] = 1 - \exp\left[-\left(\frac{x}{x_a}\right)^b\right] \tag{3.126}$$

where b is the shape parameter and x_a is the characteristic parameter. Thus the probability distribution of fatigue strength S at a given fatigue life N is derived as

$$\begin{aligned} F_S(x) &= P[S \le x] = P\left[10^{-C(\log N)^m}(S_u - S_0) + S_0 \le x\right] \\ &= P\left\{S_u \le 10^{C(\log N)^m}(x - S_0) + S_0\right\} = 1 - \exp\left\{-\left[\frac{10^{C(\log N)^m}(x - S_0) + S_0}{x_a}\right]^b\right\} \end{aligned} \tag{3.127}$$

From Eq. 3.127, it is shown that fatigue strength S at a given fatigue life N follows the three-parameter Weibull distribution.

Example 3.6 Tension-tension fatigue data for a unidirectional S2/5208 glass–epoxy laminate are given in Table 3.4. The fatigue tests were conducted using sinusoidal loading at constant load amplitude with a stress ratio of 0.1 and a

Table 3.4 Fatigue test results for $(0)_s$ S2/5208 glass–epoxy [34]

No.	Maximum stress (S/MPa)	N/cycles	No.	Maximum stress (S/MPa)	N/cycles
1	2082.54	1	15	758.54	7290
2	2048.07	1	16	758.54	6750
3	2020.48	1	17	586.15	74250
4	1979.11	1	18	586.15	67490
5	1330.90	153	19	586.15	36210
6	1289.52	267	20	586.15	49800
7	1296.42	319	21	482.71	138180
8	1344.69	436	22	482.71	93880
9	965.42	1630	23	482.71	224630
10	965.42	1330	24	482.71	55780
11	965.42	1760	25	379.27	1122310
12	965.42	1220	26	379.27	213960
13	758.54	10200	27	379.27	464810
14	758.54	9000	28	379.27	211800

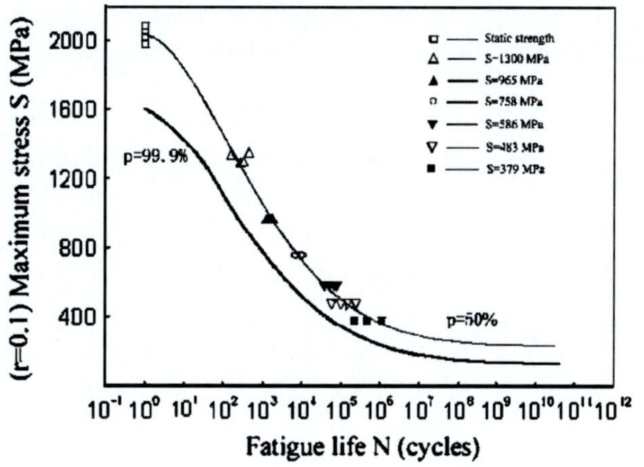

Fig. 3.16 S–N curve with four-parameter

frequency of 3 Hz. The static tests (one cycle data) were conducted at a loading rate comparable with the cyclic loading rate. The specimens came from a single panel with a V_f of 68.3%, where V_f is fibre volume fraction.

From the test data listed in Table 3.4, using the above parameter estimation method, the overall life range S–N curve equation (shown in Fig. 3.16) with four-parameter is determined as

$$\left(\frac{1795.92}{S - 236.63}\right) = 10^{5.10\times10^{-2}\times(\log N)^{1.74}} \tag{3.128}$$

Using Eq. 3.128, the mean values of fatigue life corresponding to fatigue stress $S = 965.42$, 758.54, 586.15, 482.71 and 379.27 MPa are 869, 4147, 21139, 75870 and 457457 cycles respectively. From the experimental data shown in Table 3.4, the mean values of fatigue life experimental data pertaining to fatigue stress $S = 965.42$, 758.54, 586.15, 482.71 and 379.27 MPa are 1485, 6285, 56938, 128118 and 503220 cycles respectively. The relative deviations of predicted results from experimental data are 41%, 34%, 63%, 41% and 9% respectively. According to Eq. 3.127 and the test data listed in Table 3.4, it is possible to have an S–N curve (shown in Fig. 3.16) corresponding to reliability level 99.9%.

$$\left(\frac{1568.20}{S - 236.63}\right) = 10^{0.051\times(\log N)^{1.74}} \tag{3.129}$$

Figure 3.16 shows that the calculated curves are consistent with the experimental data and the probability curves are also realistic. Thus it is argued that the overall life range S–N curve equations with four-parameter have adequately and realistically expressed the physical characteristics of overall range of fatigue damage and the varying laws of experimental data. Importantly, the parameters of this model can be estimated expediently and easily.

From Eq. 1.44, the generalized Forman formula of fatigue crack growth rate covering the overall range with four parameters is

$$\frac{da}{dN} = C\left[\left(\frac{1-f_0}{1-R}\right)\Delta K\right]^{m_1} \frac{\left(1-\frac{\Delta K_{\text{th}}}{\Delta K}\right)^{m_2}}{\left[1-\frac{\Delta K}{(1-R)K_C}\right]^{m_3}} \tag{3.130}$$

By analogy aid of Eq. 3.51, the parameters of Eq. 3.130 can be estimated. Eq. 3.130 is randomized and transformed into a natural logarithmic form as:

$$\ln\left(\frac{da}{dN}\right) = \ln C + m_1 \cdot \left[\ln\left(\frac{1-f_0}{1-R}\right) + \ln \Delta K\right] + m_2 \cdot \ln\left(1-\frac{\Delta K_{\text{th}}}{\Delta K}\right) - m_3 \cdot \ln\left[1-\frac{\Delta K}{K_C(1-R)}\right] + \ln W \tag{3.131}$$

where the random factor, W is the non-negative, stationary lognormal random variable process of state-varying with a median value of 1.0 and a standard deviation σ_w.

Letting $Y = \ln\left(\frac{da}{dN}\right)$, $X_1 = \ln\left(\frac{1-f_0}{1-R}\right) + \ln \Delta K$, $X_2 = \ln\left(1-\frac{\Delta K_{\text{th}}}{\Delta K}\right)$, $X_3 = \ln\left[1-\frac{\Delta K}{K_C(1-R)}\right]$, $a_0 = \ln C$, $a_1 = m_1$, $a_2 = m_2$, $a_3 = -m_3$, $Z = \ln W$, then Eq. 3.131 becomes

$$Y = a_0 + a_1X_1 + a_2X_2 + a_3X_3 + Z \tag{3.132}$$

where Z is the stationary lognormal random variable with a median value of 0 and a standard deviation σ_z. From Eq. 3.132, it is seen that random variable Y follows $N[a_0 + a_1x_1 + a_2x_2 + a_3x_3, \sigma_z^2]$ and the logarithm likelihood function is

$$\ln L = -n \cdot \ln \sigma_z - \frac{n}{2} \cdot \ln(2\pi) - \frac{1}{2\sigma_z^2} \cdot \sum_{i=1}^{n} (y_i - a_0 - a_1x_{1i} - a_2x_{2i} - a_3x_{3i})^2 \tag{3.133}$$

Invoking the maximum likelihood principle and using Eq. 3.133, it is possible to show

$$\sum_{i=1}^{n} y_i - na_0 - a_1\sum_{i=1}^{n} x_{1i} - a_2\sum_{i=1}^{n} x_{2i} - a_3\sum_{i=1}^{n} x_{3i} = 0 \tag{3.134}$$

$$\sum_{i=1}^{n} x_{1i}y_i - a_0\sum_{i=1}^{n} x_{1i} - a_1\sum_{i=1}^{n} x_{1i}^2 - a_2\sum_{i=1}^{n} x_{1i}x_{2i} - a_3\sum_{i=1}^{n} x_{1i}x_{3i} = 0 \tag{3.135}$$

$$\sum_{i=1}^{n} x_{2i}y_i - a_0\sum_{i=1}^{n} x_{2i} - a_1\sum_{i=1}^{n} x_{1i}x_{2i} - a_2\sum_{i=1}^{n} x_{2i}^2 - a_3\sum_{i=1}^{n} x_{2i}x_{3i} = 0 \tag{3.136}$$

$$\sum_{i=1}^{n} x_{3i}y_i - a_0\sum_{i=1}^{n} x_{3i} - a_1\sum_{i=1}^{n} x_{1i}x_{3i} - a_2\sum_{i=1}^{n} x_{2i}x_{3i} - a_3\sum_{i=1}^{n} x_{3i}^2 = 0 \quad (3.137)$$

$$n\cdot\sigma_z^2 - \sum_{i=1}^{n}(y_i - a_0 - a_1x_{1i} - a_2x_{2i} - a_3x_{3i})^2 = 0 \quad (3.138)$$

Solving Eqs. 3.134–3.137 yields

$$a_0 = \bar{y} - a_1\bar{x}_1 - a_2\bar{x}_2 - a_3\bar{x}_3 \quad (3.139)$$

$$a_1 = \frac{1}{L_{11}}(L_{10} - a_2L_{12} - a_3L_{13}) \quad (3.140)$$

$$a_2 = \frac{(L_{11}L_{20} + L_{11}L_{23} - L_{12}L_{10} - L_{12}L_{13})(L_{13}^2 - L_{11}L_{33})}{L_{11}^2L_{23}^2 + L_{11}L_{22}L_{13}^2 + L_{11}L_{12}^2L_{33} - L_{11}^2L_{22}L_{33} - 2L_{11}L_{12}L_{13}L_{23}} \quad (3.141)$$

$$a_3 = \frac{(L_{12}^2 - L_{11}L_{22})(L_{13}^2 - L_{11}L_{33}) - (L_{11}L_{20} - L_{12}L_{10})(L_{12}L_{13} - L_{11}L_{23})}{L_{11}^2L_{23}^2 + L_{11}L_{22}L_{13}^2 + L_{11}L_{12}^2L_{33} - L_{11}^2L_{22}L_{33} - 2L_{11}L_{12}L_{13}L_{23}} \quad (3.142)$$

$$\sigma_z = \sqrt{\frac{Q}{n}} \quad (3.143)$$

where

$$\bar{x}_1 = \frac{1}{n}\sum_{i=1}^{n} x_{1i}$$

$$\bar{x}_2 = \frac{1}{n}\sum_{i=1}^{n} x_{2i}$$

$$\bar{x}_3 = \frac{1}{n}\sum_{i=1}^{n} x_{3i}$$

$$\bar{y} = \frac{1}{n}\sum_{i=1}^{n} y_i$$

$$Q = \sum_{i=1}^{n}(y_i - a_0 - a_1x_{1i} - a_2x_{2i} - a_3x_{3i})^2$$

$$L_{10} = \sum_{i=1}^{n} x_{1i}y_i - n\cdot\bar{x}_1\cdot\bar{y}$$

$$L_{20} = \sum_{i=1}^{n} x_{2i}y_i - n \cdot \bar{x}_2 \cdot \bar{y}$$

$$L_{30} = \sum_{i=1}^{n} x_{3i}y_i - n \cdot \bar{x}_3 \cdot \bar{y}$$

$$L_{11} = \sum_{i=1}^{n} x_{1i}^2 - n \cdot \bar{x}_1^2$$

$$L_{22} = \sum_{i=1}^{n} x_{2i}^2 - n \cdot \bar{x}_2^2$$

$$L_{33} = \sum_{i=1}^{n} x_{3i}^2 - n \cdot \bar{x}_3^2$$

$$L_{12} = \sum_{i=1}^{n} x_{1i}x_{2i} - n \cdot \bar{x}_1 \cdot \bar{x}_2$$

$$L_{13} = \sum_{i=1}^{n} x_{1i}x_{3i} - n \cdot \bar{x}_1 \cdot \bar{x}_3$$

$$L_{23} = \sum_{i=1}^{n} x_{2i}x_{3i} - n \cdot \bar{x}_2 \cdot \bar{x}_3$$

Thus,

$$C = \exp(y_i - a_1\bar{x}_1 - a_2\bar{x}_2 - a_3\bar{x}_3) \tag{3.144}$$

$$m_1 = a_1 \tag{3.145}$$

$$m_2 = a_2 \tag{3.146}$$

$$m_3 = -a_3 \tag{3.147}$$

By analogy to Eqs. 3.49 and 3.50 and using Eq. 3.130, the probability formula of four-parameter fatigue crack growth rate curve covering the overall range is obtained.

$$\left(\frac{da}{dN}\right)_p = C\left[\left(\frac{1-f_0}{1-R}\right)\Delta K\right]^{m_1} \frac{\left(1-\frac{\Delta K_{\text{th}}}{\Delta K}\right)^{m_2}}{\left[1-\frac{\Delta K}{(1-R)K_C}\right]^{m_3}} \cdot \exp\left[u_p \hat{k} \sigma_z\right] \quad (3.148)$$

$$\left(\frac{da}{dN}\right)_{p\gamma} = C\left[\left(\frac{1-f_0}{1-R}\right)\Delta K\right]^{m_1} \frac{\left(1-\frac{\Delta K_{\text{th}}}{\Delta K}\right)^{m_2}}{\left[1-\frac{\Delta K}{(1-R)K_C}\right]^{m_3}} \cdot \exp\left\{\sigma_z \cdot \left[\hat{k}u_p + t_\gamma \sqrt{\frac{1}{n} + u_p^2\left(\hat{k}^2 - 1\right)}\right]\right\} \quad (3.149)$$

3.6 Reliability Determination of Fatigue Behaviour Based on Incomplete Data

For the same type specimens subjected to the same cyclic stress, at a specific cycle number, some are observed to fail while others run out. The cycle numbers of failed specimen is termed as the complete life data, and that of unfailed specimens is called as the incomplete life data. For a set of samples with n specimens subjected to the same cyclic stress, if r specimens failed and (n-r) specimens ran out, the complete and incomplete experimental lives can be then expressed as $(t_1, t_2, \ldots, t_r)$ and $(t_{r+1}, t_{r+2}, \ldots, t_n)$. Since the logarithmic fatigue life follows normal distribution, letting $x = \log t$, knowing x follows a normal distribution of $N(\mu, \sigma^2)$, it is possible to have probability density function of complete lives and probability of occurrence of incomplete lives respectively as follows

$$f(x_i; \mu, \sigma) = \frac{1}{\sigma\sqrt{2\pi}} e^{-\frac{(x_i-\mu)^2}{2\sigma^2}} \quad (i = 1, 2, \ldots, r) \quad (3.150)$$

$$F(x_i; \mu, \sigma) = \frac{1}{\sigma\sqrt{2\pi}} \int_{x_i}^{\infty} \exp\left[-\frac{(x-\mu)^2}{2\sigma^2}\right] dx \quad (i = r+1, \ldots, n) \quad (3.151)$$

where μ and σ are the population mean and standard deviation of normal distribution respectively. Thus, one has the likelihood function as

$$L(\mu, \sigma) = \prod_{i=1}^{r} f(x_i; \mu, \sigma) \cdot \prod_{i=1}^{n-r} \left[F(x_i; \mu, \sigma)\right]$$
$$= \prod_{i=1}^{r} \frac{1}{\sqrt{2\pi}\sigma} \exp\left[-\frac{1}{2}\left(\frac{x_i-\mu}{\sigma}\right)^2\right] \cdot \prod_{i=r+1}^{n} \frac{1}{\sqrt{2\pi}\sigma} \int_{x_i}^{\infty} \exp\left[-\left(\frac{x-\mu}{2\sigma^2}\right)^2\right] dx \quad (3.152)$$

Letting $y = \frac{x-\mu}{\sigma}$, $\Phi(x) = \int_{x}^{\infty} \exp\left[-\frac{(x-\mu)^2}{2\sigma^2}\right] dx$, then Eq. 3.152 becomes

Table 3.5 Experiments of fatigue life for helicopter rotor

No.	N/cycles	State	No.	N/cycles	State
1	411077	failed	6	1402256	failed
2	599484	failed	7	1500000	run out
3	1000238	run out	8	1956600	run out
4	1119746	failed	9	3002242	run out
5	1203120	run out	10	3191866	run out

$$L(\mu,\sigma)=\prod_{i=1}^{r}\frac{1}{\sqrt{2\pi}\sigma}\exp\left[-\frac{1}{2}y_i^2\right]\cdot\prod_{i=r+1}^{n}\frac{1}{\sqrt{2\pi}\sigma}\Phi(x_i) \tag{3.153}$$

and logarithm likelihood function is

$$\ln L(\mu,\sigma)=-\frac{n}{2}(2\ln\sigma+\ln 2+\ln\pi)-\frac{1}{2}\sum_{i=1}^{r}y_i^2+\sum_{i=r+1}^{n}\ln\Phi(x_i) \tag{3.154}$$

According to the maximum likelihood principle, from Eq. 3.154, it is possible to show

$$\frac{\partial\ln L}{\partial\mu}=\frac{1}{\sigma}\sum_{i=1}^{r}y_i+\sum_{i=r+1}^{n}\frac{\int_{x_i}^{\infty}e^{-\frac{(x-\mu)^2}{2\sigma^2}}\cdot\frac{x-\mu}{\sigma^2}dx}{\Phi(x_i)}=0 \tag{3.155}$$

$$\frac{\partial\ln L}{\partial\sigma}=-\frac{n}{\sigma}+\frac{1}{\sigma}\sum_{i=1}^{r}y_i^2+\sum_{i=r+1}^{n}\frac{\int_{x_i}^{\infty}e^{-\frac{(x-\mu)^2}{2\sigma^2}}\cdot\left(\frac{x-\mu}{\sigma}\right)^2dx}{\sigma\Phi(x_i)}=0 \tag{3.156}$$

By using an iterative method, the solutions $\hat{\mu}$ and $\hat{\sigma}$ of Eqs. 3.155 and 3.156 can be obtained. Then the safe life with a reliability level of p can be determined.

$$t_p=10^{\left(\hat{\mu}+u_p\hat{\sigma}\right)} \tag{3.157}$$

On the basis of Eq. 2.95, the one-sided low limit of fatigue life corresponding to the probability of survival of p and the confidence level of γ is

$$t_{p\gamma}=10^{\left\{\hat{\mu}+\hat{\sigma}\left[u_p-t_\gamma\sqrt{\frac{1}{n\beta^2}+u_p^2\left(1-\frac{1}{\beta^2}\right)}\right]\right\}} \tag{3.158}$$

Example 3.7 Fatigue experimental results of helicopter's composite blade are shown in Table 3.5. From Eqs. 3.155 and 3.156 and using the dataset tabled in Table 3.5, the distribution parameters of fatigue life can be estimated as

$$\hat{\mu}=6.39,\hat{\sigma}=0.502$$

From the annexed table of literature, it is possible to get

$$\mu_{0.99}=-2.326,\ t_{0.95}(v=9)=1.8331,\ \beta(n=10)=1.0280$$

Thus, according to Eq. 3.158, fatigue live pertaining to reliability levels of 50% and 99% with a confidence level of 50% are 2.45×10^6 cycles and 1.66×10^5 cycles respectively. Fatigue live pertaining to reliability levels of 50% and 99% with a confidence level of 95% are 1.28×10^6 cycles and 4.47×10^4 cycles respectively.

In order to the determine S–N curve, fatigue tests were performed at different stress levels S_i, which were chosen in the range of fatigue life of interest. At each stress level S_i, a fatigue life N_i could be obtained. If experimental data sets of $(S_i, N_i)(i = 1, 2, \ldots, n)$ consist r sets of complete test results and $(n–r)$ sets of incomplete experimental data. Thus, according to the single-point likelihood method (SPLM) proposed in Sect. 3.3, the p–S–N curve can be determined from the incomplete experimental data. Based on Eq. 1.4 the S–N curve can be described as

$$N = CS^m \tag{3.159}$$

or

$$\log N = \log C + m \cdot \log S \tag{3.160}$$

Letting $a = \log C$, $b = m$, $x = \log S$, $y = \log N$, then Eq. 3.160) becomes

$$y = a + b \cdot x \tag{3.161}$$

Similarly, it can be shown that y follows a normal distribution of $N(a + bx, \sigma^2)$, and the probability density function of complete lives and probability of occurrence of incomplete lives are respectively

$$f(x_i, y_i; a, b, \sigma) = \frac{1}{\sigma\sqrt{2\pi}} e^{-\frac{(y_i - a - bx_i)^2}{2\sigma^2}} \quad (i = 1, 2, \ldots, r) \tag{3.162}$$

$$F(x_i, y_i; a, b, \sigma) = \frac{1}{\sigma\sqrt{2\pi}} \int_{y_i}^{\infty} \exp\left[-\frac{(y - a - bx_i)^2}{2\sigma^2}\right] dy \quad (i = r + 1, \ldots, n) \tag{3.163}$$

Likelihood function is

$$\begin{aligned} L(a, b, \sigma) &= \prod_{i=1}^{r} f(x_i, y_i; a, b, \sigma) \cdot \prod_{i=r+1}^{n} F(x_i, y_i; a, b, \sigma) \\ &= \prod_{i=1}^{r} \frac{1}{\sqrt{2\pi}\sigma} \exp\left[-\frac{(y_i - a - bx_i)^2}{2\sigma^2}\right] \cdot \prod_{i=r+1}^{n} \frac{1}{\sqrt{2\pi}\sigma} \int_{y_i}^{\infty} \exp\left[-\frac{(y - a - bx_i)^2}{2\sigma^2}\right] dy \end{aligned} \tag{3.164}$$

Letting $z = \frac{y - a - bx}{\sigma}$, $\Phi(y) = \int_y^{\infty} e^{-\frac{(y - a - bx)^2}{2\sigma^2}} dy$, then logarithm likelihood function can be written as

$$\ln L(a,b,\sigma) = -\frac{n}{2}(2\ln\sigma + \ln 2 + \ln\pi) - \frac{1}{2}\sum_{i=1}^{r} z_i^2 + \sum_{i=r+1}^{n} \ln\Phi(y_i) \tag{3.165}$$

and maximum likelihood equation set are

$$\frac{\partial \ln L}{\partial a} = \frac{1}{\sigma}\sum_{i=1}^{r} z_i + \sum_{i=r+1}^{n} \frac{\int_{y_i}^{\infty} e^{-\frac{(y-a-bx_i)^2}{2\sigma^2}} \cdot \left(\frac{y-a-bx_i}{\sigma^2}\right)dy}{\Phi(y_i)} \tag{3.166}$$

$$\frac{\partial \ln L}{\partial b} = \frac{1}{\sigma}\sum_{i=1}^{r} x_i z_i + \sum_{i=r+1}^{n} \frac{x_i \cdot \int_{y_i}^{\infty} e^{-\frac{(y-a-bx_i)^2}{2\sigma^2}} \cdot \left(\frac{y-a-bx_i}{\sigma^2}\right)dy}{\Phi(y_i)} \tag{3.167}$$

$$\frac{\partial \ln L}{\partial \sigma} = -\frac{n}{\sigma} + \frac{1}{\sigma}\sum_{i=1}^{r} z_i^2 + \sum_{i=r+1}^{n} \frac{\int_{y_i}^{\infty} e^{-\frac{(y-a-bx_i)^2}{2\sigma^2}} \cdot \left(\frac{y-a-bx_i}{\sigma}\right)^2 dy}{\sigma\Phi(y_i)} \tag{3.168}$$

By using an iterative method, the solutions $\hat{a}$, $\hat{b}$ and $\hat{\sigma}$ of Eqs. 3.166–3.168 can be obtained. Then S–N and p–S–N curves can be determined respectively as

$$N = 10^{\hat{a}} S^{\hat{b}} \tag{3.169}$$

$$N_p = 10^{(\hat{a}+u_p\hat{\sigma})} S^{\hat{b}} \tag{3.170}$$

From Eq. 2.95 and 3.160, the one-sided lower limit of logarithmic fatigue life corresponding to the probability of survival of p and the confidence level of γ is

$$y_{p\gamma} = \log N_{p\gamma} = \log C + m\log S + \hat{\sigma}\left[u_p - t_\gamma\sqrt{\frac{1}{n\beta^2} + u_p^2\left(1 - \frac{1}{\beta^2}\right)}\right] \tag{3.171}$$

Thus

$$N_{p\gamma} = CS^m \cdot 10^{\left\{\hat{\sigma}\left[u_p - t_\gamma\sqrt{\frac{1}{n\beta^2} + u_p^2\left(1 - \frac{1}{\beta^2}\right)}\right]\right\}} \tag{3.172}$$

Example 3.8 This refers to a test programme for an S–N curve for a helicopter tail blade. The fatigue experimental results are listed in Table 3.6 and shown in Fig. 3.17. From Eqs. 3.166–3.168 and using the dataset tabled in Table 3.6, it is possible to obtain the parameters of the S–N curve as

$$\hat{a} = 35.61, \quad \hat{b} = -11.33, \quad \hat{\sigma} = 0.15$$

From the annexed table of literature [26], it is possible to get

$$u_{0.99} = -2.3260, \quad t_{0.95}(\upsilon = 5) = 2.0150, \quad \beta(n = 6) = 1.0510$$

Table 3.6 Experiments of S–N curve for helicopter tail rotor

No.	Moment M/N · m	N/cycles	State
1	454.09	199454	run out
2	413.11	713518	run out
3	415.14	803388	run out
4	396.38	1522000	run out
5	395.04	1449508	failed
6	412.47	545900	failed

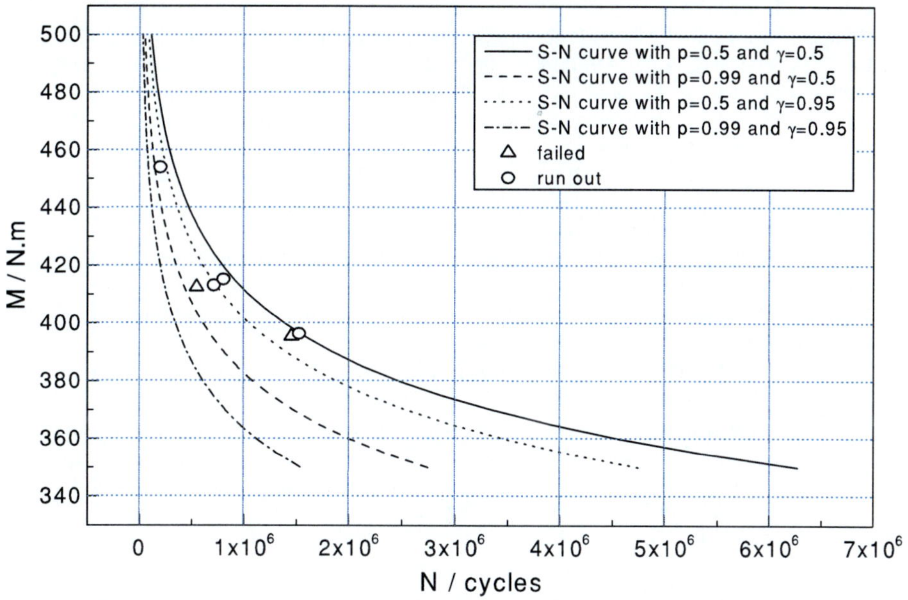

Fig. 3.17 S–N curve of helicopter tail blade

Thus, according to Eq. 3.172, the S–N curve formulae pertaining to reliability levels of 50% and 99% with a confidence level of 50% are respectively

$$N = 10^{35.61} \cdot S^{-11.33} \tag{3.173}$$

$$N = 10^{35.25} \cdot S^{-11.33} \tag{3.174}$$

The S–N curve formulae pertaining to reliability levels of 50% and 99% with a confidence level of 95% are respectively

$$N = 10^{35.49} \cdot S^{-11.33} \tag{3.175}$$

$$N = 10^{35.0} \cdot S^{-11.33} \tag{3.176}$$

Equations 3.173–3.176 are shown in Fig. 3.17. It can be seen that the calculated curves are consistent with experimental data and the probability curves again are also realistic.

References

1. Weibull W (1961) Fatigue testing and analysis of results. Macmillan Company, New York
2. Freudenthal AM, Garrelts M, Shinozuka M (1966) The analysis of structural safety. J Struct Div, ASCE 92:267–325
3. Bogdanoff JL, Kozin F (1985) Probabilistic models of cumulative damage. Wiley, New York
4. Lin YK, Wu WF, Yang JN (1985) Stochastic modeling of fatigue crack propagation: probabilistic methods in mechanics of solids and structure. Springer, Berlin
5. Proven JW (1987) Probabilistic fracture mechanics and reliability. Martinus Nijhoff, Dordrecht (The Netherlands)
6. Sobczyk K, Spencer BF (1992) Random fatigue: from data to theory. Academic Press, Boston
7. Virkler DA, Hillberry BM, Goel PK (1979) The statistical nature of fatigue crack propagation. J Eng Mater Technol, Transac ASME 101:148–153
8. Ditlevsen O, Olesen R (1986) Statistical analysis of the Virkler data on fatigue crack growth. Eng Fract Mech 25:177–195
9. Yang JN, Manning SD (1990) Stochastic crack growth analysis methodologies for metallic structures. Eng Fract Mech 37:1105–1124
10. Wu WF, Ni CC (2003) A study of stochastic fatigue crack growth modeling through experimental data. Probab Eng Mech 18(2):107–118
11. Tsurui A, Ishikawa H (1986) Application of the Fokker-Planck equation to a stochastic fatigue crack growth model. Struct Saf 4(1):15–29
12. Xiong J, Gao Z (1997) Probability distribution of fatigue damage and statistical moment of fatigue life. Science in China (Series E) 40(3):279–284
13. Yang JN, Manning SD (1996) A simple second order approximation for stochastic crack growth analysis. Eng Fract Mech 53(5):677–686
14. Ray A, Patankar R (1999) A stochastic model of fatigue crack propagation under variable-amplitude loading. Eng Fract Mech 62(4–5):477–493
15. Wu WF, Ni CC (2004) Probabilistic models of fatigue crack propagation and their experimental verification. Probab Eng Mech 19(3):247–257
16. Xiong JJ, Shenoi RA (2007) A practical randomization approach of deterministic equation to determine probabilistic fatigue and fracture behaviours based on small experimental data sets. Int J Fract 145:273–283
17. Xiong JJ, Shenoi RA (2006) Single-point likelihood method to determine a generalized S-N surface. In: Proceedings of the I Mech E (Institution of Mechanical Engineers) Part C Journal of Mechanical Engineering Science, 220(10):1519–1529
18. Xiong JJ, Shenoi RA, Zhang Y (2008) Effect of the mean strength on the endurance limit or threshold value of the crack growth curve and two-dimensional joint probability distribution. J Strain Anal Eng Des 43(4):243–257
19. Luo CY, Xiong JJ, Shenoi RA (2008) A reliability-based approach to assess fatigue behaviour based on small incomplete data sets. Adv Mater Res 44–46:871–878
20. Gallagher JP (1976) Estimating fatigue-crack lives for aircraft: techniques. Exp Mech 16:425–433
21. Rudd JL, Yang JN, Manning SD, Garver WR (1982). Durability design requirements and analysis for metallic airframes. Design of Fatigue and Fracture Resistant Structures, ASTM STP 761, 133–151

22. Artley et al. Variations in crack growth rate behaviour. 11th Fracture Mechanics Conference, ASTM STP 677, 1979:54–67
23. Lin YK, Yang JN (1983) On statistical moments of fatigue crack propagation. Eng Fract Mech 18:243–256
24. Yang JN, Donath RC (1983) Statistical fatigue crack propagation in fastener holes under spectrum loading. J Aircr 20:1028–1032
25. Yang JN, Chen S (1985) Fatigue reliability of structural components under scheduled inspection and repair maintenance, In: Probabilistic methods in the mechanics of solids and structures, Proceedings of the IUTAM symposium, Stockholm 1984, Springer, pp 559–568
26. Gao ZT, Xiong JJ (2000) Fatigue Reliability. Beihang University Press, Beijing
27. ASTM E468-76, American Society for Testing and Materials, Philadelphia, 1976
28. Nakazawa H, Kodama S (1987) Statistical S-N testing method with 14 specimens: JSME standard method for determination of S-N curves. In Statistical research on fatigue and fracture, Elsevier Applied Science, New York, 59–69
29. ASTM E 739-91 (1998) Standard practice for: statistical analysis of linearized stress-life (S-N) and strain-life (ε-N) fatigue data. In: Annual book of ASTM standards, vol 03.01, Philadelphia, 1999: 614–620
30. Ling J, Pan J (1997) A maximum likelihood method for estimating P-S-N curves. Int J Fatigue 19(5):415–419
31. Gao ZT, Xiong JJ (2000) Fatigue reliability. Beijing University of Aeronautics and Astronautics Press, Beijing, 366–372
32. Murty ASR, Gupta UC, Radha A (1995) A new approach to fatigue strength distribution for fatigue reliability evaluation. Int J Fatigue 17(2):85–89
33. Bathos C, Paris PC (2005) Gigacycle fatigue in mechanical practice. Marcel Dekker, Newyork
34. Reifsnider KL (1990) Fatigue of composite materials. Elsevier, Oxford

Chapter 4
Data Treatment and Generation of Fatigue Load Spectrum

4.1 Introduction

Fatigue reliability assessment is becoming increasingly pertinent in designing mechanical structures and components. For this, it is necessary to identify the probability density function (p.d.f.) defining the scatter in loading spectra, materials related fatigue properties data and the fatigue strength distribution function [1]. One of the fundamental issues in the reliability analysis of fatigue behaviour of real components under service loading is the evaluation of load or stress spectra and the extrapolation of complete design spectra from short term (measured or simulated) histories [2]. Because of the stochastic nature of the load (stress) ranges and histories applied to the structural component [3–9], it is usually not easy to conduct load measurements in original operating conditions. In engineering design practice, after the load history measurements are repeated several times under various operating conditions, deterministic rainflow counting procedures [10–14] that take into account the magnitude and the number of closed hysteresis loops are used since these can be performed on most kinds of stochastic processes. Additionally, the resulting load/stress time histories and corresponding loading/stress spectra are also obtained. Two major issues need to be considered in the subsequent analysis: (1) the scatter in extrapolated loading spectra from a sample of measured load time histories needs to be evaluated [2–8]; (2) the scatter in measured loading spectra needs to be determined empirically by histograms of relative frequencies [7–9].

In practice, the Palmgren–Miner linear accumulation damage rule, the S–N curve representing the material performance determined from constant-amplitude tests and load cycles defined using the rainflow algorithm are often used for fatigue life prediction of a structure subjected to a random load. Although in most cases this is the best available method, the accuracy of the approach is often quite low. Consequently, the method is generally used in the design stage, when the accuracy of fatigue life predictions is less important and when the experiments can be

J. J. Xiong and R. A. Shenoi, *Fatigue and Fracture Reliability Engineering*, Springer Series in Reliability Engineering, DOI: 10.1007/978-0-85729-218-6_4, relatively cheap. For final evaluation of fatigue life, fatigue tests on components or whole constructions under variable- amplitude loading are performed extensively. All of this testing involves generating field or service loading using lengthy complex variable amplitude histories. These histories may be generated by means of one of the following two methodologies.

(1) Using standardised spectra such as the FALSTAFF or TWIST, WASH I, load-time histories in the aircraft industry (in the design phase) are generated on the basis of a mission analysis [15–17]. However, the histories relative to the operational missions of the components and vehicles are obtained by instrumenting them and recording data in varying service conditions. As is well known [18–21], the original load-time histories often contain a large percentage of small amplitude cycles where the fatigue damage associated with these small amplitude cycles can be small. As a result, in many cases, small amplitude cycles are deleted from these histories in order to produce representative and meaningful yet economical testing. These techniques attempt to remove cycles with negligible changes to the fatigue damage or try to quantify a percentage change in the damage due to their deleting. However, deleting of variable amplitude histories can affect both crack initiation life and crack growth life and this is highly dependent upon material, history, load level and sequence [21, 22]. When generating a load history it is essential that besides parameters such as amplitude, mean and the number of load cycles, the sequence of the load cycles is taken into account. This sequence can often substantially influence a fatigue crack's growth rate.
(2) A new load history is generated on the basis of a distribution of load cycles extracted from the original load history. The methods range from classical autoregressive methods [23–25] which are used in generating uncompressed load histories to Markov methods [26, 27] which are used in generating compressed load histories. A common weakness of these methods is that the generation of new load histories is not based on the rainflow counting method, the method most often used for extracting load cycles when an estimation of the correlation between dynamic loads and the structure's fatigue life is performed. An advantage of the rainflow method is not only that the load cycles extracted from a complex, random sequence with this method correspond to the closed hysteresis loops in the diagram but it also that it enables the sequence of load cycles to be taken into consideration when calculating the fatigue damage. However, these methods have two weaknesses. Firstly, the generated load history can be composed only of those load cycles that were extracted from the original load history and secondly, the information about the sequence of load cycles is lost sometimes.

The problem is that none of the above-mentioned methods alone fulfils all the requirements that ought to be considered when treating original random load histories and generating new random load histories. These requirements are as follows:

- What kind of a counting method should be employed for the dataset treatment of divergence–convergence waves?
- How should the scatter in extrapolated loading spectra from a sample of measured load time histories be expressed and how should the minimum sampling numbers be determined?
- How should the distribution type be tested and how should the necessary estimation parameters be established?
- How should the information about the sequence of load cycles be maintained and how could it be ensured that the damage resulting from the new generating load history is likely to be the same as that from the original load history?
- How should the new generated load histories be of shorter length than the original load history to decrease the test time through deleting the small load cycles and merging the smaller ones?

It is desirable to have a systematic procedure to address the dataset treatment of original load spectrum and to develop a new method for the generation of load histories, which will address all these requirements in assessing structural reliability. Therefore, in this chapter, a modified counting method for the dataset treatment of divergence–convergence waves and a new method for the scatter prediction of loading spectra are proposed. Further, a load history generation approach is established for full-scale accelerated fatigue tests through deleting small amplitude carrier cycles and merging a certain number of secondary cycles.

4.2 Rain Flow-Loop Line Scheme

A prerequisite for service life determination is the extrapolation and evaluation of fatigue load spectra of real components under service loading as shown in Fig. 4.1. It is seen that the data treatment system to extrapolate and evaluate fatigue load spectra has the following functions: (1) extracting all load cycles from the load history, (2) determining the lowest number for sampling, (3) estimating the parameters of one- and two-dimensional distributions, (4) testing the distribution type. This data treatment system could be integrated for more efficient, practical fatigue load spectrum obtainment.

Though a great variety of variations of the rainflow counting method have been developed, the method developed by Matsuishi–Endo [26] is still regarded as one of the most practical and valid in recognizing damaging events in a complex loading history. It is based on the cyclic material behaviour that can be described by a simple mathematical model consisting of two modules, namely 'Masing behaviour' and 'memory rules'. The classical rainflow cycle counting is illustrated by the example given in Fig. 4.2. The stress-time history is shown in Fig. 4.2a. This leads to the stress–strain path in Fig. 4.2b that forms the closed hysteresis loops shown in Fig. 4.2c. The sequence of the reversals 1 to 7 is registered in the order (2–3–2′), (5–6–5′), (1–4–7). An equivalent form is (2–3), (5–6), (1–4)

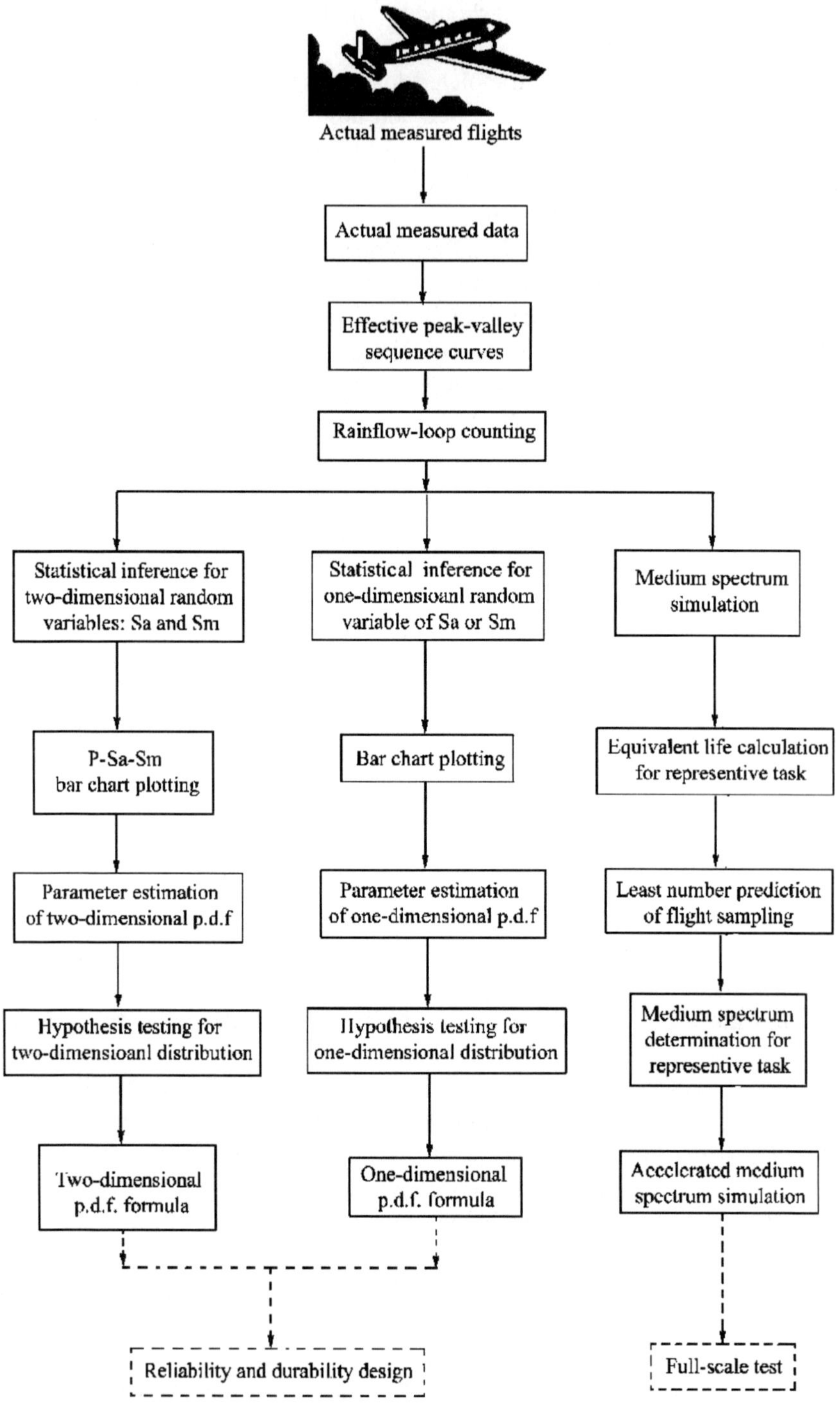

Fig. 4.1 Data treatment system for original load history

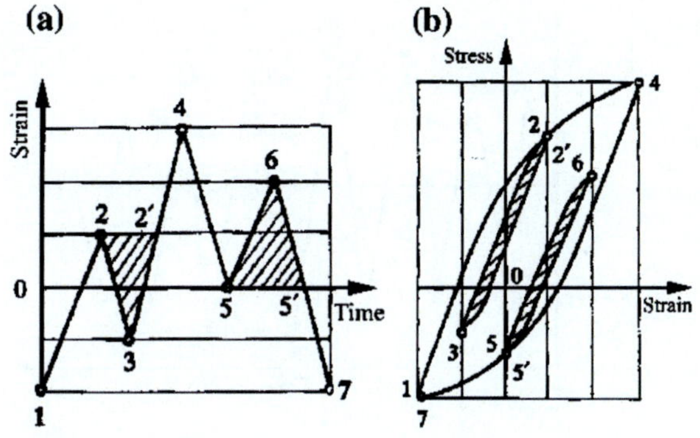

Fig. 4.2 Example of the classical rainflow cycle counting. (**a**) Stress-time history. (**b**) Hysteresis loops. (**c**) Counted load cycles

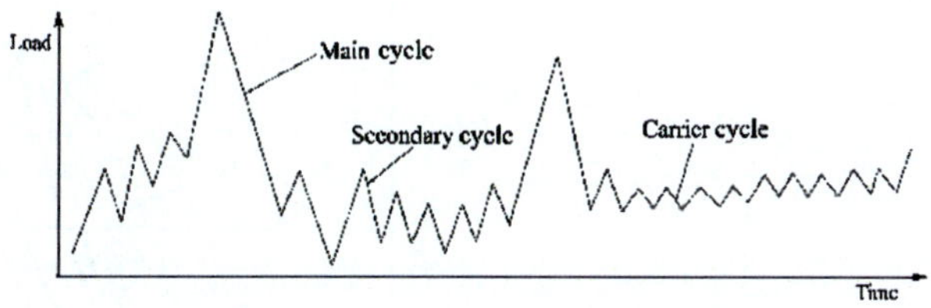

Fig. 4.3 An original load history

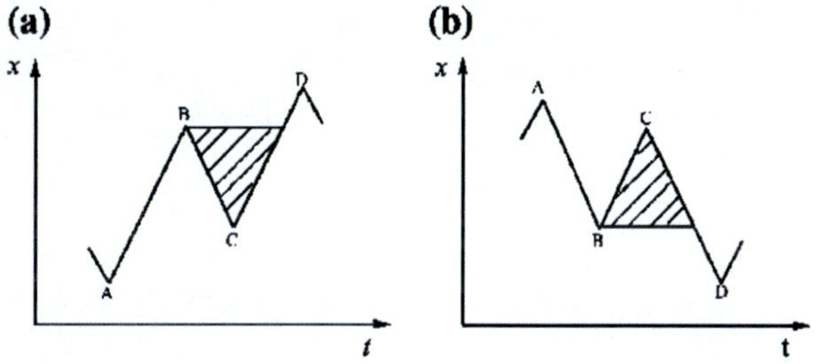

Fig. 4.4 Representative load cycle. (**a**) Hanging load cycle. (**b**) Standing load cycle

because the start and end points of the closed loop are not differentiated when damage is determined.

From an original load-time history (shown in Fig. 4.3), it is seen that there are generally two types of representative load cycle waves (shown in Fig. 4.4) and four consecutive points A, B, C and D fulfill the following relations:

$$|x_C - x_B| \leq |x_B - x_A| \tag{4.1}$$

$$|x_C - x_B| \leq |x_D - x_C| \tag{4.2}$$

By using the rainflow counting method, it is possible to extract the shadowing load cycles (see Fig. 4.4) from the load histories. The mean and amplitude of individual extracted load cycle are respectively:

$$s_a = \frac{|x_C - x_B|}{2} \tag{4.3}$$

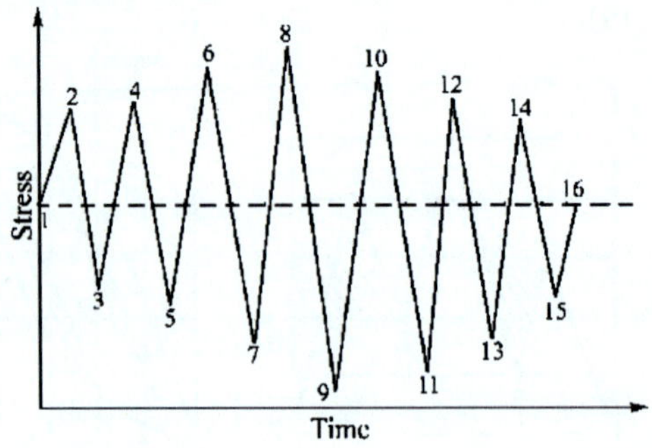

Fig. 4.5 Divergence–convergence wave

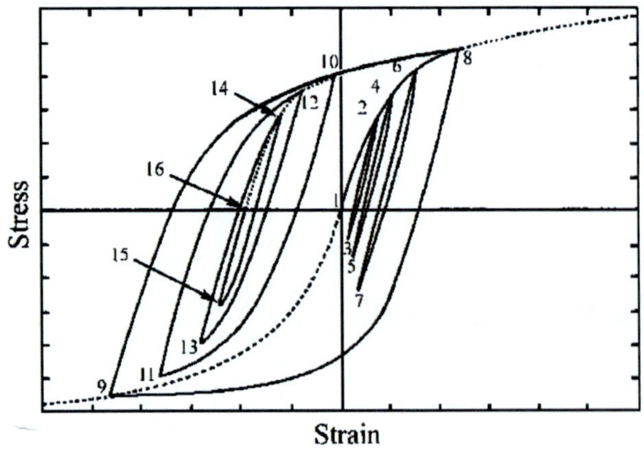

Fig. 4.6 Hysteresis loops of convergence–divergence wave

$$s_{\rm m} = \frac{x_{\rm C} + x_{\rm B}}{2} \tag{4.4}$$

From the above illustration, it is clear the load identification rules (4.1) and (4.2) and load extracting formulas (4.3) and (4.4) can be employed expeditiously for computation coding. Using the classical rainflow cycle counting mentioned above to extract the load cycles from an original load-time history (shown in Fig. 4.3), a final divergence–convergence wave-like load history (shown in Fig. 4.5) remains, for which there exist the following relations [24]:

$$s_{i+2} \geq s_i \tag{4.5}$$

and:

$$s_{i+1} \leq s_{i-1} \tag{4.6}$$

The hysteresis loops for a divergence–convergence wave-like load history as shown in Fig. 4.5 are illustrated in Fig. 4.6. It is seen from Fig. 4.6 that there exist the open loops of 1–8–9–10, 10–11–12, 12–13–14, 14–15–16 and the close loops of 2–3, 4–5, 6–7 when convergence–divergence counting is conducted; so the load

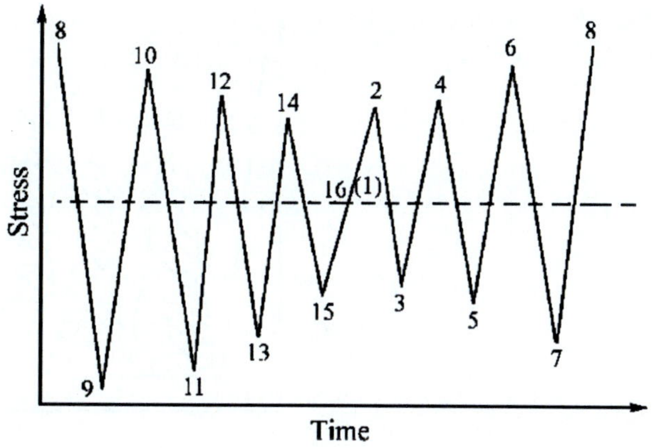

Fig. 4.7 Convergence–divergence wave

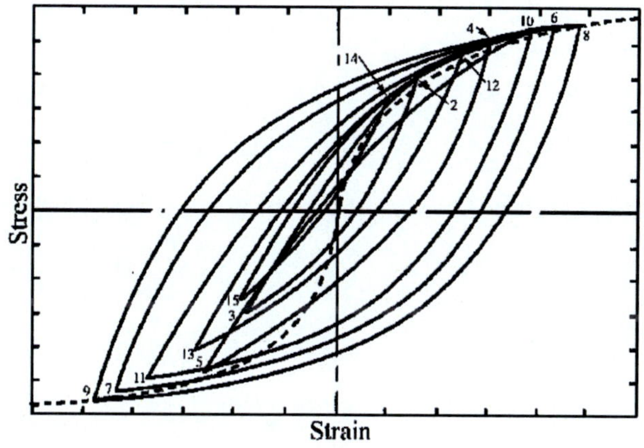

Fig. 4.8 Hysteresis loops of convergence–divergence wave

cycles of 2–3, 4–5, 6–7 can be extracted while those of 1–8–9–10, 10–11–12, 12–13–14, 14–15–16 cannot. The modification of the classical rainflow method presented now is very simple but decisive. The divergence–convergence wave-like load history is separated into two stages at the highest peak (such as point 8 in Fig. 4.5) or lowest valley (such as point 9 in Fig. 4.5) and linking the first point with the last point. This constitutes a convergence–divergence wave-like load history as shown in Fig. 4.7. Using the classical rainflow counting again, all hysteresis loops of this load history can be closed, as shown in Fig. 4.8, and all load cycles can be extracted from this load history, as shown in Figs. 4.9 and 4.10. The counting method mentioned above for the divergence–convergence wave-like load history is termed as the loop-based counting method due to its development from the stress–strain hysteresis loop.

The above loop-based counting method can be allowed to transform a divergence–convergence wave (shown in Fig. 4.11) into a convergence–divergence wave-like load history (shown in Fig. 4.12). For four consecutive points $j = i - 1, \ldots, i + 2$ as shown in Figs. 4.11 and 4.12, it is possible to have the following relations:

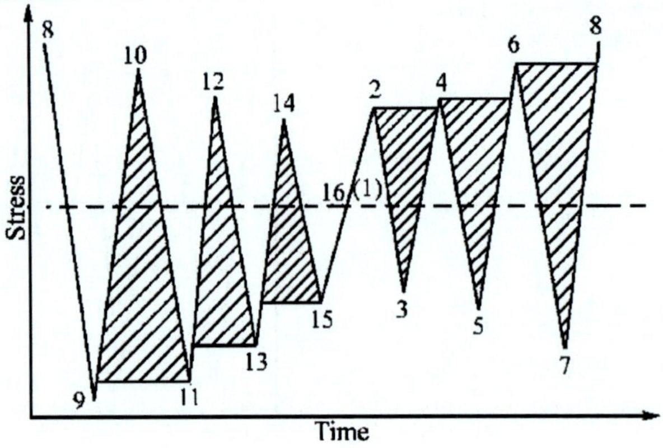

Fig. 4.9 Counted load cycles

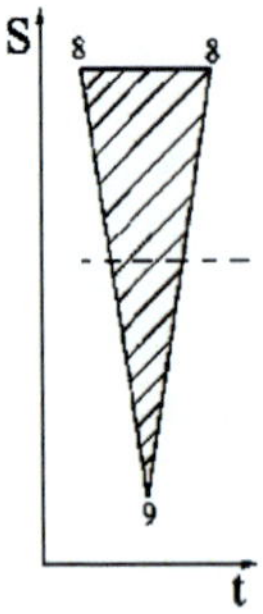

Fig. 4.10 Last load cycle counted

$$(s_{i-1} - s_{i+1})(s_{i+2} - s_i) > 0 \tag{4.7}$$

$$s_i - s_{i+1} > 0 \tag{4.8}$$

Thus the amplitude and mean of the load cycle extracted with the convergence–divergence counting then are respectively [24]:

$$s_a = \frac{|s_{i+1} - s_i|}{2} \tag{4.9}$$

$$s_m = \frac{s_{i+1} + s_i}{2} \tag{4.10}$$

where the stress s_{i+1} corresponding to reversal point $i + 1$ being a half-value in the load cycle. In this way the load cycle is completely defined with a pair of reversal stress values (s_i, s_{i+1}) from the divergence–convergence wave-like load history. It is obvious that all of the load cycles (see Figs. 4.13 and 4.14) can be extracted from an original divergence–convergence wave-like load history (see Fig. 4.11) through the convergence–divergence wave-like load history (see Fig. 4.12) using the convergence–divergence counting procedure mentioned above.

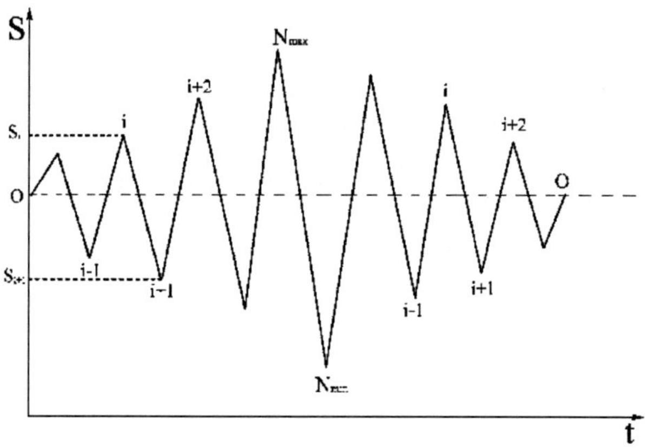

Fig. 4.11 Divergence–convergence wave

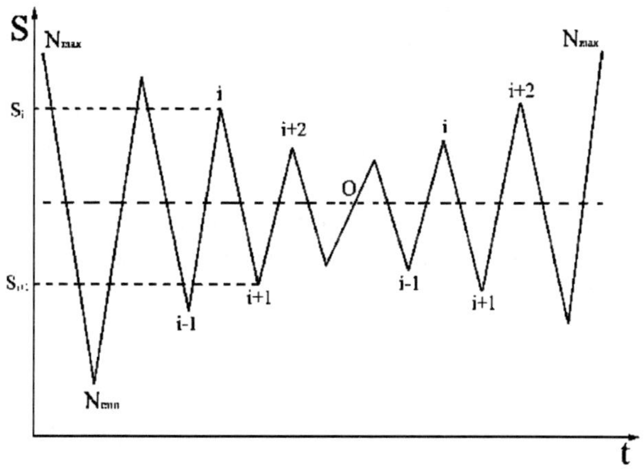

Fig. 4.12 Convergence–divergence wave

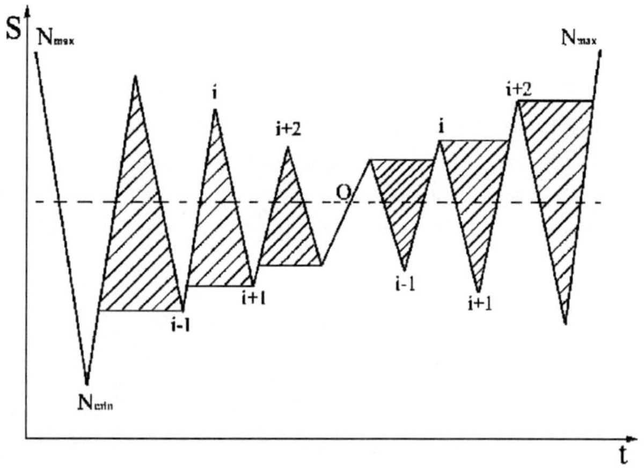

Fig. 4.13 Counted load cycles

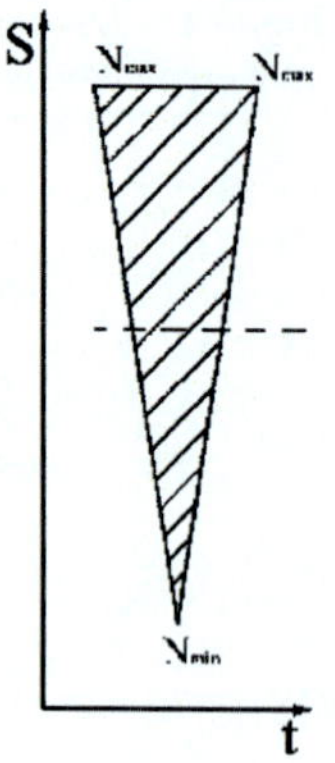

Fig. 4.14 Last load cycle counted

4.3 Two-Dimensional Probability Distribution of Fatigue Load

After the original load spectra are measured, it is necessary to use a proper probability density function to describe of the full range of cycles, in order to analyse the data obtained and to evaluate the occurrence frequency of ranges in the analysed loading conditions. In general, two components of the stress cycle, namely the amplitude s_a and the mean s_m, are assumed to be random variables based on original measured data. Their optimal distribution types are obtained by assessing the relative merits of the Gauss and Weibull distributions in treating the original measured data by hypothesis testing. The probability density functions of Gauss and Weibull distributions are respectively:

$$f(s_a) = \frac{1}{\sigma\sqrt{2\pi}} e^{-\frac{(s_a-\mu)^2}{2\sigma^2}} \tag{4.11}$$

$$f(s_a) = \frac{b}{A_a - A_0}\left(\frac{s_a - A_0}{A_a - A_0}\right)^{b-1} \exp\left[-\left(\frac{s_a - A_0}{A_a - A_0}\right)^{b}\right] \tag{4.12}$$

where μ and σ are the population mean value and standard deviation of random variable s_a respective. A_0 is the characteristic or minimum, A_a is the scale parameter and b is the shape parameter.

Recent research [24] shows that for some mechanical components, their original load spectrum follows the semi-Gauss distribution, not the Gauss and Weibull distributions. The probability density function of semi-Gauss distribution is:

$$f(s_a) = \frac{\sqrt{2}}{\sigma_1\sqrt{\pi}} \, e^{-\frac{(s_a-\mu_1)^2}{2\sigma_1^2}}, \qquad (\mu_1 \le s_a < \infty) \tag{4.13}$$

where μ_1 and σ_1 are the population parameters, but not the population mean and standard deviation. As is shown in Fig. 4.15, the estimator $\hat{\mu}_1$ of population parameter μ_1 is the abscissa value of point A, or the minimum value of stress amplitude.

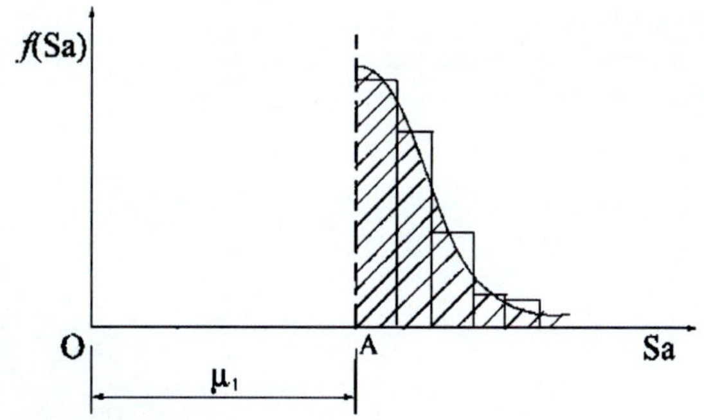

Fig. 4.15 P.d.f curve of the semi-Gauss

Taking the vertical line through point A as the symmetry axis, assuming that the same symmetric bar chart exists on the left side of symmetry axis as that on the right, the estimator $\hat{\mu}_1$ and $\hat{\sigma}_1$ of population parameter μ_1 and σ_1 can be determined by the following equation as:

$$\begin{cases} \hat{\mu}_1 = \min\{s_{a1}, s_{a2}, \ldots, s_{an}\} \\ \hat{\sigma}_1 = \sqrt{\dfrac{2\sum\limits_{i=1}^{n}(s_{ai} - \hat{\mu}_1)^2}{2n-1}} \end{cases} \tag{4.14}$$

Hypothesis testing results demonstrate that fatigue stress amplitude s_a generally follows a semi-Gauss or a three parameter Weibull distribution, whereas fatigue stress mean s_m follows the three parameter Weibull or Gauss distributions. In case where stress amplitude s_a follows a Weibull distribution with three parameters, mean stress s_m follows a Gauss distribution and s_a and s_m are mutually independent statistics, then from Eqs. 4.11 and 4.12, it is possible to obtain the two-dimensional p.d.f. of load cycle as:

$$f(s_a, s_m) = \frac{b}{A_a - A_0} \cdot \left(\frac{s_a - A_0}{A_a - A_0}\right)^{b-1} \cdot \frac{1}{\sigma\sqrt{2\pi}} \cdot \exp\left\{-\left[\left(\frac{s_a - A_0}{A_a - A_0}\right)^b + \frac{(s_m - \mu)^2}{2\sigma^2}\right]\right\} \tag{4.15}$$

The distribution given in Eq. 4.15 is illustrated in Fig. 4.16.

In case that stress amplitude s_a follows a semi-Gauss distribution, mean stress s_m follows a Gauss distribution and s_a and s_m are mutually independent statistics, then from Eqs. 4.11 and 4.13, one has the two-dimensional p.d.f of load cycle as:

$$f(s_a, s_m) = \frac{1}{\pi\sigma_1\sigma_2} \cdot \exp\left\{-\frac{1}{2}\left[\left(\frac{s_a - \mu_1}{\sigma_1}\right)^2 + \left(\frac{s_m - \mu_2}{\sigma_2}\right)^2\right]\right\} \tag{4.16}$$

where μ_1 and σ_1 are the population parameters. μ_2 and σ_2 are the population mean and standard deviation respectively. Equation 4.16 is shown in Fig. 4.17.

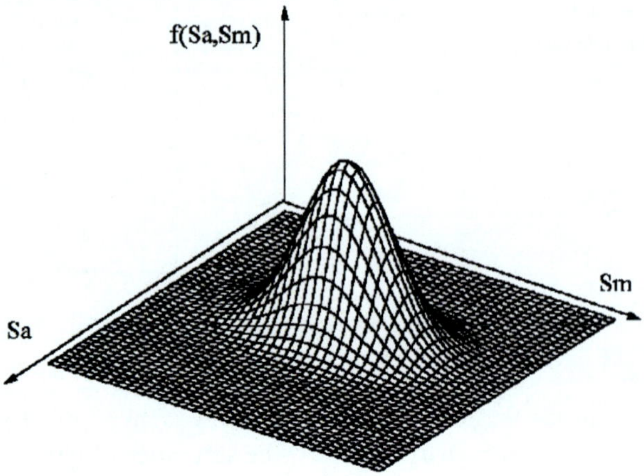

Fig. 4.16 Combined p.d.f surface of the Weibull and Gauss distributions

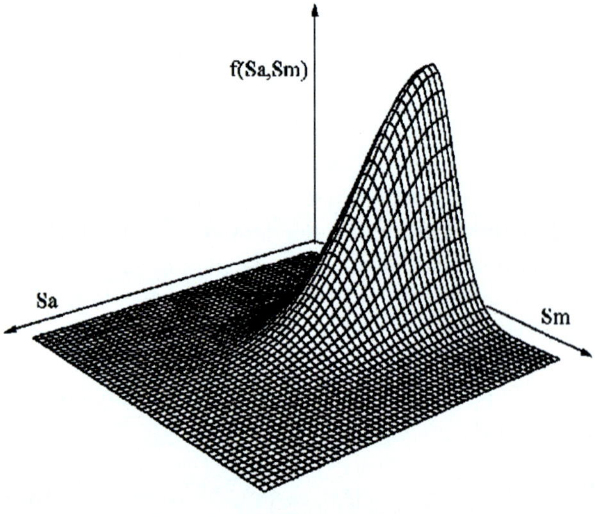

Fig. 4.17 Combined p.d.f surface of the semi-Gauss and Gauss distributions

4.4 Quantification Criteria to Identify Load Cycle

As mentioned in Sect. 4.2, all load cycles can be extracted from an original load history using the rainflow-loop count method. These extracted load cycles can be classified into three categories of load cycles, namely main, secondary and carrier cycles.

Main cycles are the few larger load cycles, which probably cause significant damage to a structure; thus should be maintained during the new load history generation. An original history often contains a large percentage of small amplitude cycles, the fatigue damage associated with which can be small. These small amplitude and high frequency load cycles are termed as carrier cycles. As a result, in many cases, carrier cycles are deleted from the original history in order to

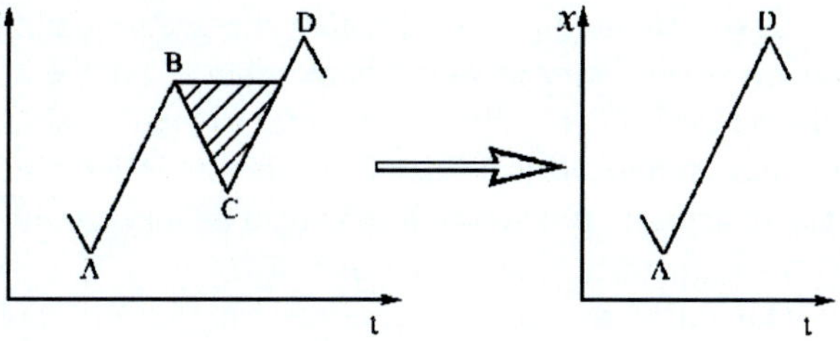

Fig. 4.18 Deleted carrier cycle from original load history

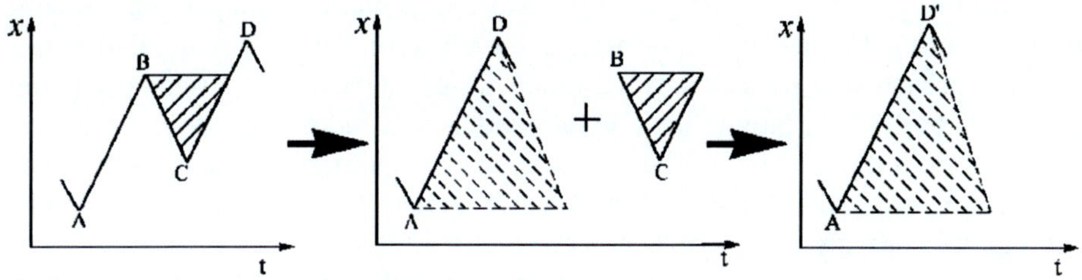

Fig. 4.19 A new secondary cycle merged from two adjacent and sequential secondary cycles

produce a meaningful yet economical testing load history (shown in Fig. 4.18). The procedure proposed to delete a carrier cycle is as follows [25]:

(1) Based on the rainflow counting, two adjacent and sequential cycles (A, D) and (B, C) can be extracted from an original load history A-B-C-D.
(2) According to the criterion of deleting small loads, the carrier load cycle (B, C) can be deleted; the remainder load cycle is (A, D). This remainder cycle (A, D) is used to replace the two original secondary cycles (A, D) and (B, C).
(3) The new load history A-D is used to replace the original load history A-B-C-D.

As mentioned above, it is possible to define a secondary cycle as one larger than a carrier cycle but smaller than a main cycle. Because of the great numbers of secondary cycles in any original load history, adjacent and sequential secondary cycles should be merged into a new secondary cycle (shown in Fig. 4.19). From Fig. 4.19, one can deduce a procedure to merge the adjacent and sequential secondary cycles as below [25].

(1) Based on the rainflow counting procedure, two adjacent and sequential cycles (A, D) and (B, C) can be extracted from an original load history A-B-C-D.
(2) According to the equivalent damage formulations given below, in case of the minimum stress of the new secondary cycle being equal to or less than one of the two adjacent and sequential secondary cycles, a new secondary cycle (A, D′) can be determined from above two secondary cycles (A, D) and (B, C). This new secondary cycle (A, D′) is used to replace the two original secondary cycles (A, D) and (B, C); the merging of two adjacent and sequential secondary cycles (A, D) and (B, C) is thus accomplished.
(3) The new load history A–D′ is used to replace the original load history A-B-C-D.

It is worth noting that if the new merged secondary cycle becomes a main cycle then it is necessary to stop this merging. Thus the information about the sequence of load cycles is maintained in order to retain a similar interaction effect. Further, the new generated load history is shorter in length than the original load history. The generation process of a new load history can be represented schematically by a block diagram as shown in Fig. 4.20.

Obviously, in a newly generated load history, only main and secondary cycles associated with fatigue damage need to be taken into account. Actually, the classification of the three categories of load cycles varies with fatigue damage growth. During fatigue crack formation, for instance, where structural damage shows slip band and micro-crack initiation inside the material, stress cycles of magnitude lower than fatigue limit should not cause structural damage. These are called as the carrier cycles. The stress cycles pertaining to stable slow crack growth rate are termed as the secondary cycles and those pertinent to unstable rapid growth rate are main cycles.

According to fatigue mechanics, in general, for low cycle fatigue life (less than $N_{\mathrm{T}} = 10^4$ cycles), the stress level is so large that the structure under this stress enters plastic deformation; here structural damage is not capable of being predicted exactly by means of the S–N curve. Hence, this kind of stress cycle corresponding to low cycle fatigue life (less than N_{T} cycles) is regarded as a main cycle. Usually, for longer lives (more than $N_{\mathrm{f}} = 10^7$ cycles), the stress level is less than fatigue limit pertaining to N_{f} cycles and should not cause structural damage. For a mid-range life (more than N_{T} cycles and less than N_{f} cycles), the stress level is likely to be in the elastic range. Consequently, structural damage is capable of being predicted by means of the S–N curve. Further, the stress cycle pertaining to this mid-range life is determined as a secondary cycle. Therefore, from Eq. 1.18 and above mentioned definitions of three categories of load cycles, the quantification criteria to identify load cycle for load history generation of fatigue accelerated tests can be established as:

$$\sigma_{\mathrm{b}} s_{\mathrm{a}} + \left[\left(\frac{C}{N_{\mathrm{f}}}\right)^{\frac{1}{m}} + S_0\right] s_{\mathrm{m}} \leq \left[\left(\frac{C}{N_{\mathrm{f}}}\right)^{\frac{1}{m}} + S_0\right] \sigma_{\mathrm{b}} \tag{4.17}$$

$$\sigma_{\mathrm{b}} s_{\mathrm{a}} + \left[\left(\frac{C}{N_{\mathrm{T}}}\right)^{\frac{1}{m}} + S_0\right] s_{\mathrm{m}} \geq \left[\left(\frac{C}{N_{\mathrm{T}}}\right)^{\frac{1}{m}} + S_0\right] \sigma_{\mathrm{b}} \tag{4.18}$$

If the load cycle extracted by Eqs. 4.1–4.4 satisfies the inequality (4.17), then this load cycle can be identified as a carrier cycle. If the load cycle extracted by Eqs. 4.1–4.4 satisfies the inequality (4.18), then this load cycle is as a main cycle. If a load cycle does not satisfy either of inequalities (4.17) and (4.18), then it can be regarded as a secondary cycle.

A large number of experimental results pertaining to fatigue crack growth rate for metallic materials have been shown to reveal that the stable or linear crack growth rate region (i.e. Region 2) generally includes a range between

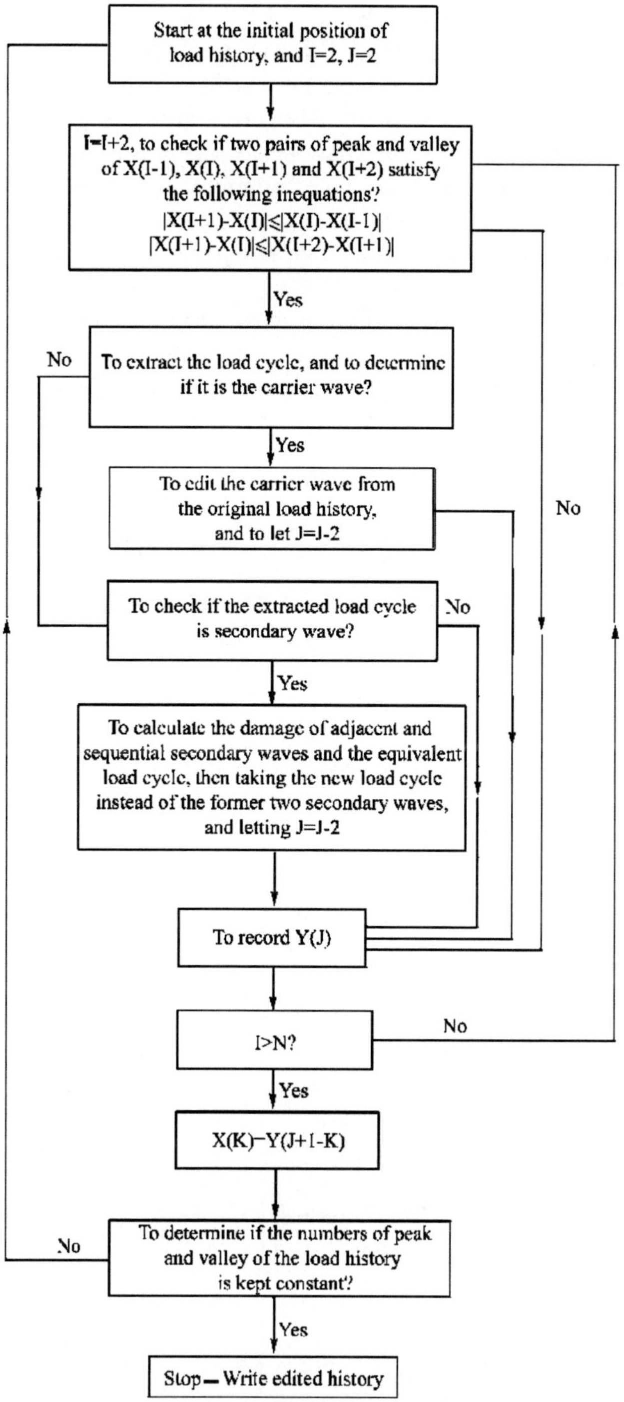

Fig. 4.20 A block diagram of the load history generation for accelerated tests

$(da/dN)_f = 10^{-6}$ mm/cycle and $(da/dN)_T = 10^{-4}$ mm/cycle (shown in Fig. 1.16). The lower part of crack growth curve with a slope of less than $(da/dN)_f$ is termed as the near-threshold region (or Region 1), while the upper part with a slope of grater than $(da/dN)_T$ is called as the unstable crack growth region (i.e. Region 3). Though it is well-known that short crack behaviour in Region 1 also affects the fatigue life, the influence of a stress cycle pertaining to a lower crack growth rate than $(da/dN)_f$ in Region 1 on fatigue life is much smaller than those in Regions 2 and 3 to be neglected from an engineering viewpoint. Therefore, the stress cycle pertinent to Region 1 is regarded in the following as a carrier cycle and is deleted from the original load history according to above mentioned deleting procedure of carrier cycles. Furthermore, the stress cycles corresponding to Regions 2 and 3 are deemed as secondary and main cycles respectively and need to be maintained. From Eq. 1.42, fatigue crack growth rate (the Walker formula) is expressed as:

$$da/dN = C(\Delta K)^{m_1}(1-R)^{m_2}$$

where R is the stress ratio. m_1 and m_2 are the exponents of fatigue $da/dN - \Delta K$ formula.

Substituting Eq. 1.51 into above equation and simplifying, it is possible to have the Walker formulae with and without plastic zone correction at crack tip respectively:

$$\frac{da}{dN} = C\left\{\frac{s_m + s_a}{\sqrt{1-\pi\alpha[(s_m+s_a)/\sigma_s]^2}} - \frac{s_m - s_a}{\sqrt{1-\pi\alpha[(s_m-s_a)/\sigma_s]^2}}\right\}^{m_1}[Y(a)]^{m_1}\left(\frac{2s_a}{s_m+s_a}\right)^{m_2} \tag{4.19}$$

$$\frac{da}{dN} = 2^{(m_1+m_2)}C\cdot s_a^{(m_1+m_2)}(s_m+s_a)^{-m_2}[Y(a)]^{m_1} \tag{4.20}$$

where $Y(a) = \alpha(a)\sqrt{\pi a}$; $\alpha(a)$ is the fatigue crack geometric correction coefficient. a_{cr} is the critical crack size. Therefore, from Eq. 4.19 and the above mentioned definitions of three categories of load cycles, the quantification criteria to identify load cycle for load history generation of accelerated fracture tests with plastic zone correction at crack tip can be obtained as:

$$\left\{\frac{s_m + s_a}{\sqrt{\sigma_s^2-\pi\alpha(s_m+s_a)^2}} - \frac{s_m - s_a}{\sqrt{\sigma_s^2-\pi\alpha(s_m-s_a)^2}}\right\}^{m_1}\left(\frac{s_a}{s_m+s_a}\right)^{m_2} \le \frac{(da/dN)_f}{2^{m_2}C[\sigma_s\cdot Y(a_0)]^{m_1}} \tag{4.21}$$

$$\left\{\frac{s_m + s_a}{\sqrt{\sigma_s^2-\pi\alpha(s_m+s_a)^2}} - \frac{s_m - s_a}{\sqrt{\sigma_s^2-\pi\alpha(s_m-s_a)^2}}\right\}^{m_1}\left(\frac{s_a}{s_m+s_a}\right)^{m_2} \ge \frac{(da/dN)_T}{2^{m_2}C[\sigma_s\cdot Y(a_0)]^{m_1}} \tag{4.22}$$

where a_0 is the initial crack size. From an engineering viewpoint, a_0 can be generally chosen as a visible and detectable macro-crack size of, say, about 1.00 mm, a value chosen on the basis of practical considerations. If the load cycle extracted by Eqs. 4.1–4.4 satisfies the inequality (4.21), then this load cycle can be identified as a carrier cycle. If the load cycle extracted by Eqs. 4.1–4.4 satisfies the inequality (4.22), then this load cycle is as a main cycle. If a load cycle does not satisfy either of inequalities (4.21) and (4.22), then it can be regarded as a secondary cycle.

Similarly, from Eq. 4.20 and the above mentioned definitions of three categories of load cycles, it is possible to have the quantification criteria to identify load cycle for load history generation of fracture accelerated tests without plastic zone correction at crack tip:

$$\frac{s_a^{(m_1+m_2)}}{(s_m+s_a)^{m_2}} \le \frac{(da/dN)_f}{2^{(m_1+m_2)}C[Y(a_0)]^{m_1}} \tag{4.23}$$

$$\frac{s_a^{(m_1+m_2)}}{(s_m+s_a)^{m_2}} \ge \frac{(da/dN)_T}{2^{(m_1+m_2)}C[Y(a_0)]^{m_1}} \tag{4.24}$$

Equations 4.23 and 4.24 are simpler forms of Eqs. 4.21 and 4.22 and therefore easier to be utilized for the load history generation, as shall be seen in the experimental verification examples and application in full-scale fracture test of a helicopter tail structure later in the chapter.

4.5 Equivalent Damage Formulations

As an original history often contains a large percentage of secondary cycles, adjacent and sequential secondary cycles should be merged into a new secondary cycle to shorten test time based on the damage equivalence principle. According to the Palmgren–Miner rule, the damage $D(S_a, S_m)$ resulting from a stress cycle of (S_a, S_m) is:

$$D(s_a, s_m) = \frac{1}{N(s_a, s_m)} \tag{4.25}$$

where $N(s_a, s_m)$ is determined by using Eq. 1.18. From Eqs. 4.25 and 1.18, the equivalent damage of a new merged stress cycle comprising two adjacent and sequential secondary cycles of (S_{a1}, S_{m1}) and (S_{a2}, S_{m2}) shown in Fig. 4.19 can be determined as:

$$\left[\frac{\sigma_b}{\sigma_b-(s_m)_{eq}}(s_a)_{eq}-S_0\right]^{-m} = \left(\frac{\sigma_b}{\sigma_b-s_{m1}}s_{a1}-S_0\right)^{-m} + \left(\frac{\sigma_b}{\sigma_b-s_{m2}}s_{a2}-S_0\right)^{-m} \tag{4.26}$$

Letting $(s_{\min})_{eq} = s_{\min 2} = s_{m2} - s_{a2}$, $M = \left[\left(\frac{\sigma_b}{\sigma_b - s_{m1}} s_{a1} - S_0\right)^{-m} + \left(\frac{\sigma_b}{\sigma_b - s_{m2}} s_{a2} - S_0\right)^{-m}\right]^{-\frac{1}{m}} + S_0$,

Then from Eq. 4.26, the stress amplitude and mean of the new merged secondary cycle are respectively:

$$(s_a)_{eq} = \frac{(\sigma_b + s_{a2} - s_{m2})M}{\sigma_b + M} \tag{4.27}$$

and:

$$(s_m)_{eq} = \frac{\sigma_b(s_{m2} - s_{a2} - M) + 2M(s_{m2} - s_{a2})}{\sigma_b + M} \tag{4.28}$$

Separating the variables of Eq. 4.20 and integrating, the crack growth life under the stress cycle (S_a, S_m) of a secondary cycle in the elastic range can be expressed as:

$$N = \frac{(s_m + s_a)^{m_2}}{2^{(m_1+m_2)} C s_a^{(m_1+m_2)}} \int_{a_0}^{a_{cr}} [Y(a)]^{-m_1} da \tag{4.29}$$

where $Y(a) = \alpha(a)\sqrt{\pi a}$. $\alpha(a)$ is the fatigue crack geometric correction coefficient. a_{cr} is the critical crack size, which can be determined using the following equation as:

$$a_{cr} \cdot \alpha(a_{cr}) = \frac{K_C}{(s_m + s_a)\sqrt{\pi}} \tag{4.30}$$

where K_c is the fracture toughness.

Letting:

$$Q = \frac{\int_{a_0}^{a_{cr}} a^{\frac{-m_1}{2}} [\alpha(a)]^{-m_1} da}{2^{(m_1+m_2)} \pi^{\frac{m_1}{2}} C} \tag{4.31}$$

then Eq. 4.29 becomes:

$$N = Q \cdot \frac{(s_m + s_a)^{m_2}}{s_a^{(m_1+m_2)}} \tag{4.32}$$

Using Eq. 4.26 as an analogy, it is possible to have equivalent damage formula of fatigue crack growth as:

$$\frac{(s_a)_{eq}^{(m_1+m_2)}}{\left[(s_m)_{eq} + (s_a)_{eq}\right]^{m_2}} = \frac{s_{a1}^{(m_1+m_2)}}{(s_{m1} + s_{a1})^{m_2}} + \frac{s_{a2}^{(m_1+m_2)}}{(s_{m2} + s_{a2})^{m_2}} \tag{4.33}$$

Letting:

$$(s_{\min})_{eq} = s_{\min 2} = s_{m2} - s_{a2}, \quad H = \frac{s_{a1}^{(m_1+m_2)}}{(s_{m1}+s_{a1})^{m_2}} + \frac{s_{a2}^{(m_1+m_2)}}{(s_{m2}+s_{a2})^{m_2}}$$

then, from Eq. 4.33, then one has:

$$(s_a)_{eq} = \left[H \cdot (s_{m2}+s_{a2})^{m_2}\right]^{\frac{1}{m_1+m_2}} \tag{4.34}$$

and:

$$(s_m)_{eq} = s_{m2} - s_{a2} + \left[H \cdot (s_{m2}+s_{a2})^{m_2}\right]^{\frac{1}{m_1+m_2}} \tag{4.35}$$

4.6 Experimental Verification

In order to verify the criteria for load cycle identification and the generation principles for accelerated load histories, three kinds of specimens made of LY12 aluminum alloy, 40CrNiMoA and 30CrMnSiNi2A alloyed steels were used in fatigue comparative tests between the original and accelerated load histories.

4.6.1 Test 4.1

Concerns LY12 aluminum alloy specimens of a shape and size shown in Fig. 4.21. The test is carried out on an MTS-880-50KN fatigue testing machine at loading frequency of 15 Hz under room temperature and atmospheric conditions to verify the generation approach of load history for accelerated test. The original load history is shown in Fig. 4.22. From the literature [28], the generalized fatigue S–N surface (stress concentration factor $K_t = 2.5$) and $da/dN - \Delta K$ curve of LY12 aluminum alloy are respectively:

$$N = 3.86 \times 10^9 \left(\frac{460.0}{460.0 - S_m} S_a - 27.0\right)^{-2.43} \tag{4.36}$$

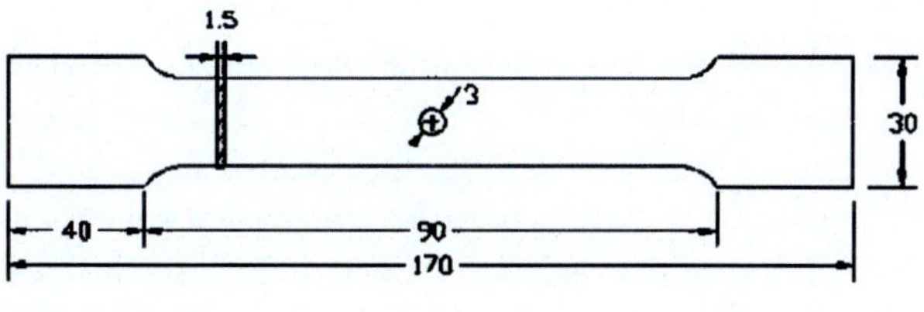

Fig. 4.21 The specimen (size unit: mm)

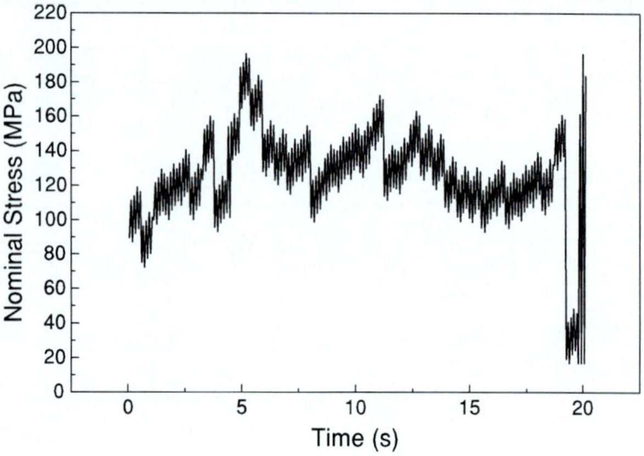

Fig. 4.22 Original load history

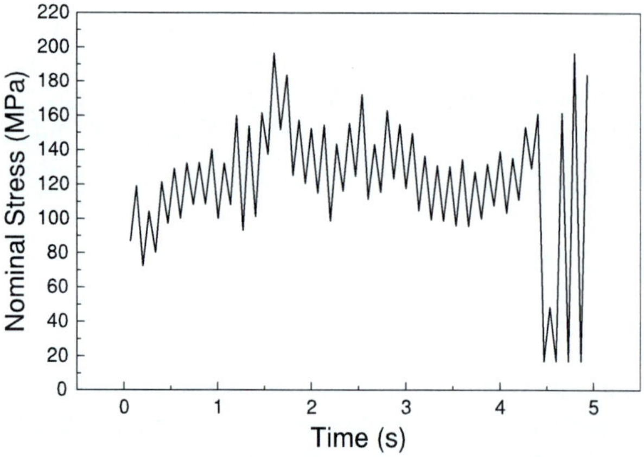

Fig. 4.23 Generation load history for accelerated tests

$$\frac{da}{dN} = 1.19 \times 10^{-2}(\Delta K)^{3.83}(1 - R)^{-1.43} \tag{4.37}$$

with $\Delta K_{th} = 2.8\,\mathrm{MPa} \cdot \sqrt{\mathrm{m}}$ at the stress ratio of $R = 0.05$.

By means of the accelerated load history generation methodology proposed earlier, a new accelerated load history is generated (shown in Fig. 4.23). In Figs. 4.22 and 4.23, all carrier cycles are deleted, the secondary cycles are merged largely, and the main cycle and the sequence between main and secondary cycles are maintained. From Figs. 4.22 and 4.23, it can be deduced that there are 151 load cycles in a block of the original load history, and 37 cycles in a block of the accelerated test load history. Five and eight specimens are used for fatigue testing using the original and accelerated load histories respectively, and experimental

Table 4.1 Experimental lives of LY12 aluminum alloy specimens

Specimen No.	Original load history		Generated load history	
	(cycles)	(blocks)	(cycles)	(blocks)
1	249603	1653	58940	1592
2	227708	1508	52133	1409
3	256851	1701	57830	1562
4	226500	1500	71891	1943
5	206568	1368	59681	1613
6			56425	1525
7			56017	1513
8			59239	1501
Mean life		1546		1594.75
Total test time	21 h 37 min		8 h 45 min	
Mean test time	4 h 19 min		1 h 5 min	

results are shown in Table 4.1. From Table 4.1, it is found that the total test time for five specimens under the original load history is 1297 min (or 21 h and 37 min) and the mean test time for every specimen is 259 min (or 4 h and 19 min). The total test time for eight specimens under the accelerated load history is 525 min (namely 8 h and 45 min) and the mean test time for every specimen is 65.6 min (or 1 h and 5 min). The relative deviation in the mean life of the accelerated test from the original load history test is about $\frac{|1594.75-1546|}{1594.75} \times 100\% = 3.15\%$, and the average saved test time for every specimen is 193.4 min (3 h and 13 min). This implies adequately close agreement between the shortened or accelerated test programme and the original or extended test programme for engineering application.

4.6.2 *Test 4.2*

Concerns 40CrNiMoA alloyed steel specimens of a shape and size shown in Fig. 4.24. All specimens have an initial prefabricated crack of 0.5 mm through linear cutting and polishing. The tests are carried out on an MTS-880-500KN fatigue testing machine at a loading frequency of 10 Hz under room temperature and atmospheric pressure conditions. As in the previous test, the original load history is shown in Fig. 4.25. The $da/dN - \Delta K$ curve of 40CrNiMoA alloyed steel at the stress ratio of $R = 0.1$ is obtained as:

$$\frac{da}{dN} = 1.56 \times 10^{-4}(\Delta K)^{2.95} \tag{4.38}$$

with $\Delta K_{\text{th}} = 5.54\,\text{MPa} \cdot \sqrt{m}$.

Again, using the same method as in Test 1, a new accelerated load history is obtained (shown in Fig. 4.26). From Figs. 4.25 and 4.26, it can be deduced that there are 609 load cycles in a block of the original load history, and 139 cycles in a

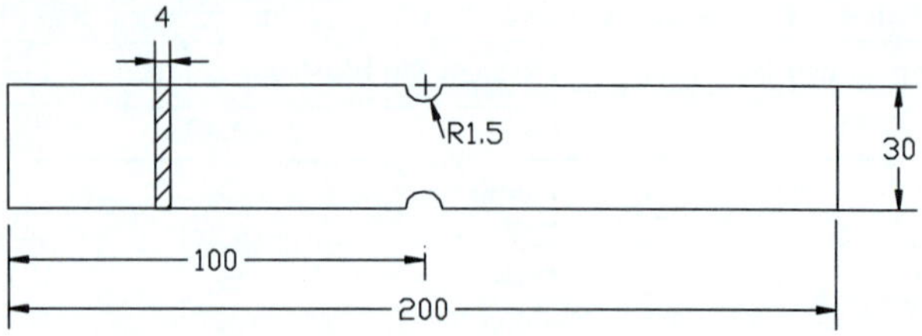

Fig. 4.24 The specimen (size unit: mm)

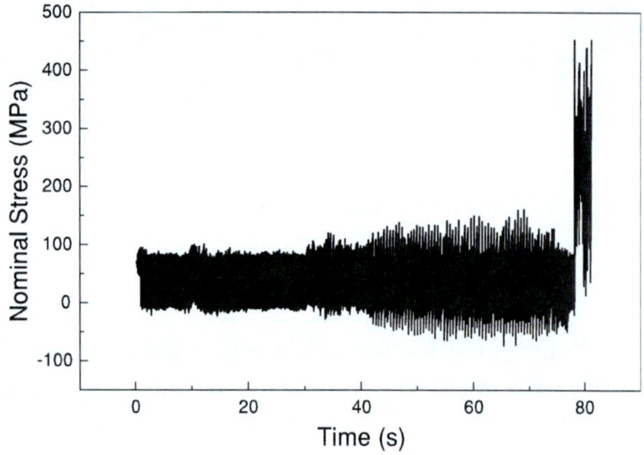

Fig. 4.25 Original load history

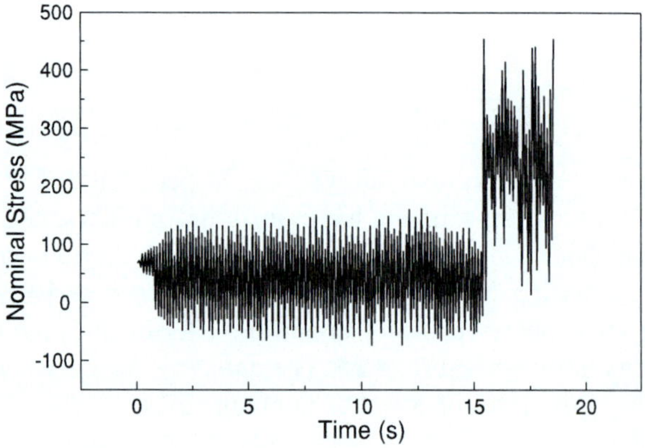

Fig. 4.26 Generation load history for accelerated tests

block of the accelerated test load history. Six and nine specimens are used for fatigue testing using the original and accelerated load histories respectively, and experimental results are shown in Table 4.2. From Table 4.2, it is found that total

Table 4.2 Experimental lives of 40CrNiMoA alloyed steel specimens

Specimen No.	Original load history		Generated load history	
	(cycles)	(blocks)	(cycles)	(blocks)
1	1091937	1793	213173	1533
2	834330	1370	203770	1465
3	984758	1617	168739	1213
4	841635	1382	216563	1558
5	878765	1442	169031	1216
6	760587	1247	206972	1489
7			194739	1401
8			220988	1589
9			177781	1279
Mean life		1475.17		1415.89
Total test time	149 h 46.7 min		49 h 13 min	
Mean test time	24 h 57.8 min		5 h 28 min	

test time for six specimens under the original load history is 8986.69 min (or 149 h and 46.7 min) and the mean test time for every specimen is 1497.78 min (or 24 h and 57.8 min). The total test time for nine specimens under the accelerated load history is 2952.9 min (or 49 h and 13 min) and the mean test time for every specimen is 328.1 min (or 5 h and 28 min). The relative deviation in the mean life of the accelerated crack propagation test from the original load history test is about $\frac{|1475.17-1415.89|}{1475.17} \times 100\% = 4.02\%$, and the average saved test time for every specimen is 1169.68 min (19 h and 30 min). Again, it is evident that the new approach using accelerated test data gives results that agree well with the original test programme for engineering application.

4.6.3 Test 4.3

Concerns 30CrMnSiNi2A alloyed steel specimens of a shape and size shown in Fig. 4.27. The tests are carried out on an MTS-880-500KN fatigue testing machine at a loading frequency of 10 Hz under room temperature and atmospheric pressure

Fig. 4.27 The specimen (size unit: mm)

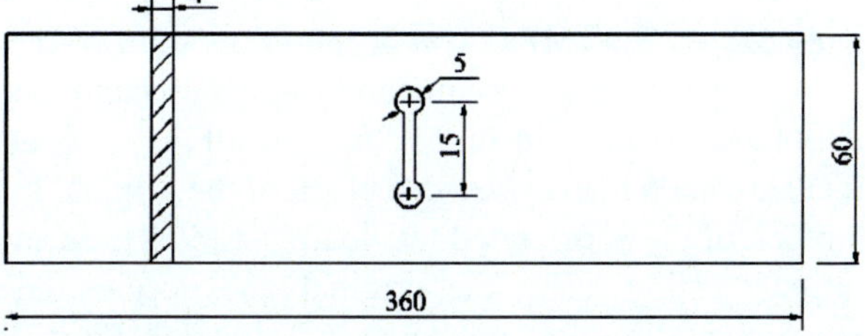

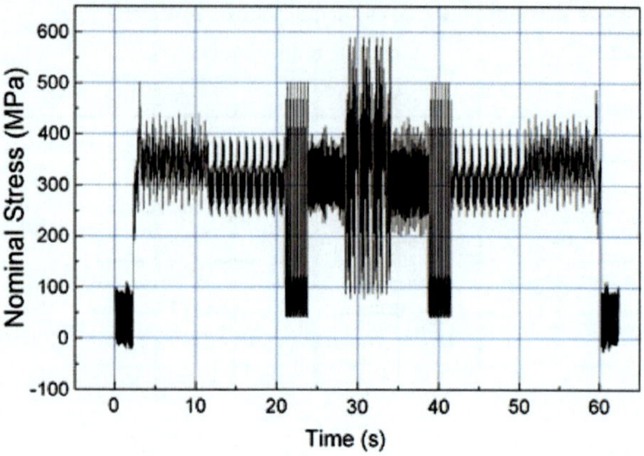

Fig. 4.28 Original load history

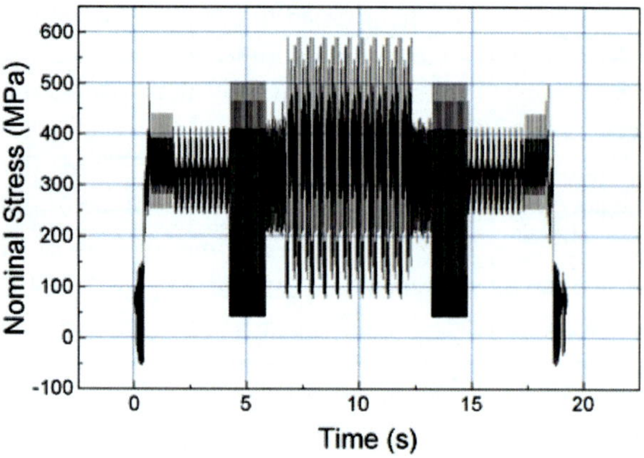

Fig. 4.29 Generation load history for accelerated tests

conditions. The original load history is shown in Fig. 4.28. The $da/dN - \Delta K$ curve of 30CrMnSiNi2A alloyed steel is obtained as:

$$\frac{da}{dN} = 3.011 \times 10^{-8} (\Delta K)^{2.45} (1 - R)^{-0.98} \quad (4.39)$$

with $\Delta K_{\mathrm{th}} = 3.67\,\mathrm{MPa} \cdot \sqrt{m}$ at the stress ratio of $R = 0.1$.

Again, using the same method as for Tests and 2, a new accelerated load history is obtained (shown in Fig. 4.29). From Figs. 4.28 and 4.29, it can be shown that there are 6906 load cycles in a block of the original load history, and 1663 cycles in a block of the accelerated test load history. Three and five specimens are used for fatigue testing using the original and accelerated load histories respectively; experimental results are shown in Table 4.3. From Table 4.3, it is calculated that the total test time for the three specimens under the original load spectra is

Table 4.3 Experimental lives of 30CrMnSiNi2A alloyed steel specimens

Specimen No.	Original load history		Generated load history	
	(cycles)	(blocks)	(cycles)	(blocks)
1	367340	53	87307	50
2	297193	43	65194	37
3	427556	61	115307	66
4			106088	60
5			62739	35
Mean life		52.3		53.25
Mean test time	10 h 6 min		2 h 1 min	

1820 min (or 30 h and 20 min) and the mean test time for every specimen is 607 min (or 10 h and 6 min). The total test time for the five specimens under the accelerated load spectra is 728 min (or 12 h and 8 min) and the mean test time for every specimen is 121 min (or 2 h and 1 min). From Table 4.3, it is evident that the relative deviations in the mean value of fatigue life under the accelerated spectra compared to that under the original spectrum test is $\frac{|53.25-52.31|}{53.25} \times 100\% = 1.82\%$. The saved test time for every specimen is about 451 min (7 h and 31 min). Again, it is evident that the new approach using accelerated test data gives results that agree well with the original test programme for engineering application.

4.7 Application in Full-Scale Fatigue Test of Helicopter Tail

A full-scale fatigue test was carried out on a helicopter tail (shown in Fig. 4.30) at loading frequency of 15 Hz under room temperature and atmospheric conditions. Fatigue load spectra in transverse and vertical directions were applied to the tail rotor as shown in Fig. 4.31 to model its original flight loads. The critical sections and hazardous locations can be identified from the original measured load spectra

Fig. 4.30 Full-scale fatigue test of helicopter tail

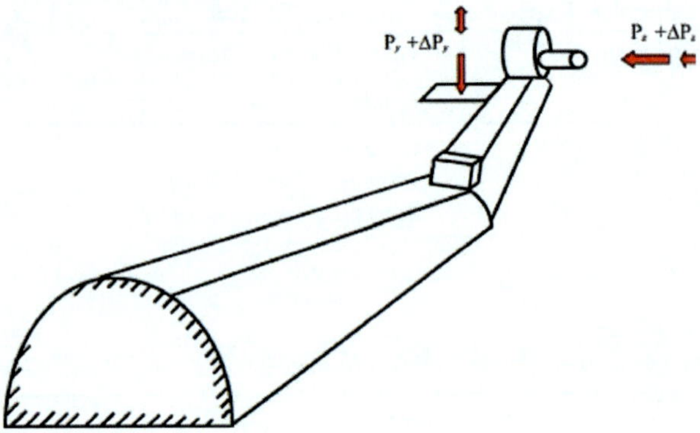

Fig. 4.31 Fatigue load application points for full-scale test of helicopter tail

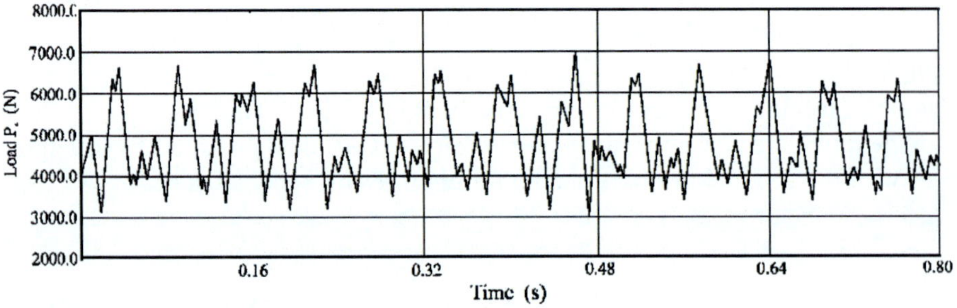

Fig. 4.32 Fatigue load spectrum in transverse direction

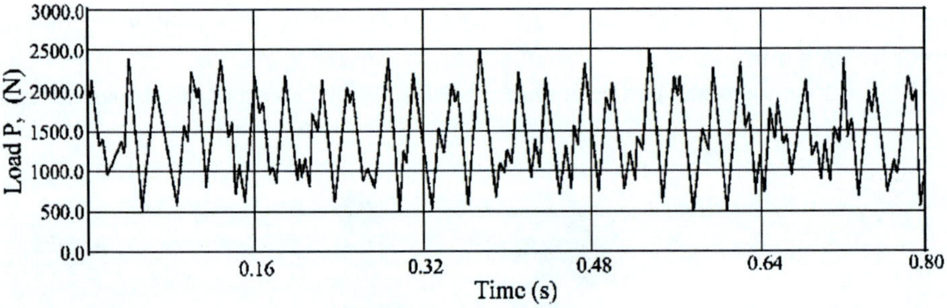

Fig. 4.33 Fatigue load spectrum in vertical direction

and using the finite element method (FEM) to determine stress hot spots. Subsequently, by means of Eqs. 4.1–4.4, 4.23, 4.24 and fracture performances of material in potentially hazardous locations, all load cycles can be extracted from the original measured load history and identified into the carrier, secondary and main cycles. According to the procedure shown in Fig. 4.20, the carrier cycles

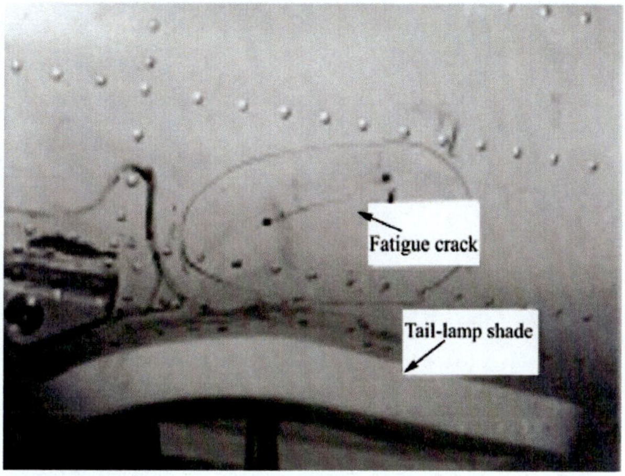

Fig. 4.34 Fatigue crack occurrence site

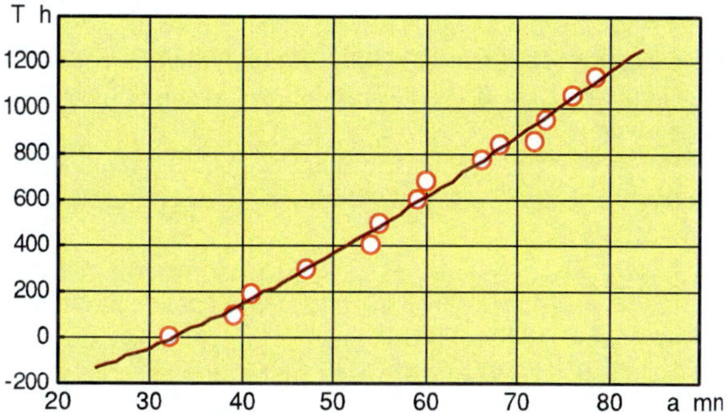

Fig. 4.35 Fatigue crack growth curve

were deleted, the secondary cycles were merged based on Eqs. 4.34, 4.35 and fracture performances of material in hazardous locations assessed. Then the new load histories for accelerated test in transverse and vertical directions were generated as shown in Figs. 4.32 and 4.33. During the fatigue tests, responses at all critical sections and in potentially hazardous locations were monitored at various test times to find fatigue crack occurrence. After 10512 blocks of load history (or 22506 flight hours), an S-shape crack with a length of 32 mm appears at the skin surface of bottom helicopter tail near the right region of tail-lamp shade (shown in Fig. 4.34). The resultant fatigue crack growth curve is shown in Fig. 4.35. The full-scale fatigue test has spend an experimental time of 3658.9 h (or 11253 blocks of load history or 22506 flight hours) until the crack reaches a length of 84.5 mm.

References

1. Jeon WS, Song JH (2002) An expert system for estimation of fatigue properties of metallic materials. Int J Fatigue 24(6):685–698
2. Tovo R (2000) A damage-based evaluation of probability density distribution for rain-flow ranges from random processes. Int J Fatigue 22(5):425–429
3. Tovo R (2002) Cycle distribution and fatigue damage under broad-band random loading. Int J Fatigue 24(11):1137–1147
4. Nagode M, Fajdiga M (1998) On a new method for prediction of the scatter of loading spectra. Int J Fatigue 20(4):271–277
5. Klemenc J, Fajdiga M (2000) Description of statistical dependencies of parameters of random load states (dependency of random load parameters). Int J Fatigue 22(5):357–367
6. Nagode M, Klemence J, Fajdiga M (2001) Parametric modelling and scatter prediction of rainflow matrices. Int J Fatigue 23:525–532
7. Nagode M, Fajdiga M (1998) A general multi-modal probability density function suitable for the rainflow ranges of stationary random processes. Int J Fatigue 20(3):211–223
8. Olagnon M (1994) Practical computation of statistical properties of rainflow counts. Int J Fatigue 16:306–314
9. Zhao W, Baker MJ (1992) On the probability density function of rainflow stress range for stationary Gaussian processes. Int J Fatigue 14(2):121–135
10. Amzallag C, Gerey JP, Robert JL, Bahuaud J (1994) Standardization of the rainflow counting method for fatigue analysis. Int J Fatigue 16:287–293
11. Downing SD, Socie DF (1982) Simple rainflow counting algorithms. Int J Fatigue 4:31–40
12. Glinka G, Kam ICP (1987) Rainflow counting algorithm for very long stress histories. Int J Fatigue 9:223–228
13. Hong N (1991) A modified rainflow counting method. Int J Fatigue 13:465–469
14. Rychlik I (1987) A new definition of the rainflow cycle counting method. Int J Fatigue 9(2):119–121
15. Fowler KR, Watanabe RT (1989) Development of jet transport airframe fatigue test spectra. In: Potter JM, Watanabe RT (eds) Development of fatigue loading spectra. ASTM STP 1006, pp 36–64. ASTM, Philadelphia
16. Schütz W (1989) Standardized stress-time histories-An overview. In: Potter JM, Watanabe RT (eds) Development of fatigue loading spectra. ASTM-STP 1006, pp 3–16. ASTM, Philadelphia
17. Heuler P, Klätschke H (2005) Generation and use of standardised load spectra and load-time histories. Int J Fatigue 27(8):974–990
18. Schijve J, Vlutters AM, Ichsan A, Kluit JCP (1985) Crack growth in aluminium alloy sheet material under flight-simulation loading. Int J Fatigue 7(3):127–136
19. Yan JH, Zheng XL, Zhao K (2001) Experimental investigation on the small-load-omitting criterion. Int J Fatigue 23(5):403–415
20. Schön J (2006) Spectrum fatigue loading of composite bolted joints-Small cycle elimination. Int J Fatigue 28(1):73–78
21. Socie DF, Artwohl PJ (1980) Effect of history editing on fatigue crack initiation and propagation in a notched member. In: Bryan DF, Potter IM (eds) Effect of load history variables on fatigue crack initiation and propagation. ASTM STP 714, pp 3–23. ASTM, Philadelphia
22. Pompetzki TH, Topper TH, DuQuesnay DL (1990) The effect of compressive underloads and tensile overloads on fatigue damage accumulation in SAE 1045 steel. Int J Fatigue 12:207–213
23. Klemenc J, Fajdiga M (2004) An improvement to the methods for estimating the statistical dependencies of the parameters of random load states. Int J Fatigue 26(2):141–154
24. Xiong JJ, Shenoi RA (2005) An integrated and practical reliability-based data treatment system for actual load history. Fatigue Fract Eng Mater Struct 28(10):875–889

25. Xiong JJ, Shenoi RA (2008) A load history generation approach for full-scale accelerated fatigue tests. Eng Fract Mech 75(10):3226–3243
26. Matsuishi M, Endo T (1968) Fatigue of metals subjected to varying stress. In: Proceedings of the Kyushu Branch of Japan Society of Mechanics Engineering, Fukuoka, Japan (in Japanese). 1968, pp 37–40
27. Anthes RJ (1997) Modified rainflow counting keeping the load sequence. Int J Fatigue 19(7):529–535
28. Gao ZT, Jiang XT, Xiong JJ, Guo GH, Gan WM, Xia QY, Wang SP, Zeng BY (1999) Test design and data treatment for fatigue behaviour. Beihang University Press, Beijing

Chapter 5
Reliability Design and Assessment for Total Structural Life

5.1 Introduction

It has been reported that 80–90% of failures in steel structures are related to fatigue and fracture [1, 2]. Therefore, fatigue reliability analyses now are widely used because of the requirement of safe operation of mechanical structures. When a component has to endure rather low stress levels and very long stress cycles, it can be designed according to the infinite life design method of high cycle fatigue. For example, vehicle suspension components, axles, or crankshafts of engines may be required to sustain more than 10^8 load cycles. When a component such as an airplane's structure or heavy-duty machinery, etc. works under high and low stress cycles, it can be designed on the basis of the finite life design method of low cycle fatigue. From the viewpoint of fatigue reliability, "infinite life design" means that the component does not crack or if it has cracked already then the crack no longer grows during an ultra-long life. Certain reliability and confidence levels can be ascribed to this event. A reduction factor method generally is implemented to determine safe fatigue endurance limit S_R, or safe fracture threshold. This implies that in the case of maximum fatigue stress $S_{\max} > S_R$, failure will occur in a component having endured the appropriate number of cycles at the given stress level. Conversely in the case of $S_{\max} < S_R$, then damage will not appear in a component subjected the cyclic stress of infinite cycles [3–6]. This kind of design principle is suitable for the case of steady low stress levels and high cycles, by controlling fatigue stress to be below safe fatigue limit, or by assuring that the stress intensity factor is below the safe facture threshold. A one-dimensional stress-strength interference model was initially applied to fatigue reliability analysis by Freudenthal et al. [7]. Later, the model was extensively used in ultra-long life fatigue reliability analyses [8–11]. Here, "one-dimensional stress-strength" implies that fatigue stress and fatigue strength are described by their amplitude value with mean value being regarded as a constant. In fact,

J. J. Xiong and R. A. Shenoi, *Fatigue and Fracture Reliability Engineering*, Springer Series in Reliability Engineering, DOI: 10.1007/978-0-85729-218-6_5, however fatigue life and fatigue damage are controlled by two-dimensional random variables (stress amplitude S_a and mean stress S_m). Hence, two-dimensional stress-strength interference models were developed [12, 13].

According to the finite life design method of low cycle fatigue, safe structural life can be determined from experiments or theoretical calculations by using scatter factor formulae [6, 14]. Up to now, in order to guard against failures from unforeseen circumstances, two major approaches to structural substantiation have been devised. One is fatigue analysis through a testing programme that attempts to establish a 'safe life' for the structure under assumed loading conditions. This procedure implies that life can be predicted and that the structure can be inspected, repaired and restored or retired from service in this predicted life time. If the analysis can be established fairly early on in the design process, then any deficiencies can be eliminated or minimised. It has also been recognised that inevitably some structural damage and failures would occur and that catastrophic failure is almost never tolerable. This has led to approaches that are termed 'damage tolerant' or 'fail safe' designs, in which the damage would be temporarily tolerated until repair can be effected or the damage assumes potentially critical dimensions. Both approaches are of interest and have been complementary to each other. Consequently, in the chapter, a series of structural reliability design and assessment methodologies are revealed and discussed.

5.2 Probability Method for Infinite Life Design

If x_1 and x_2 represent two mutually independent normal random variables pertinent to fatigue stress and fatigue limit respectively and their probability density functions (p.d.f) (shown in Fig. 5.1) are respectively as:

$$f(x_1) = \frac{1}{\sigma_1\sqrt{2\pi}} e^{-\frac{(x_1-\mu_1)^2}{2\sigma_1^2}} \tag{5.1}$$

$$g(x_2) = \frac{1}{\sigma_2\sqrt{2\pi}} e^{-\frac{(x_2-\mu_2)^2}{2\sigma_2^2}} \tag{5.2}$$

Fig. 5.1 One-dimensional stress-strength interference model

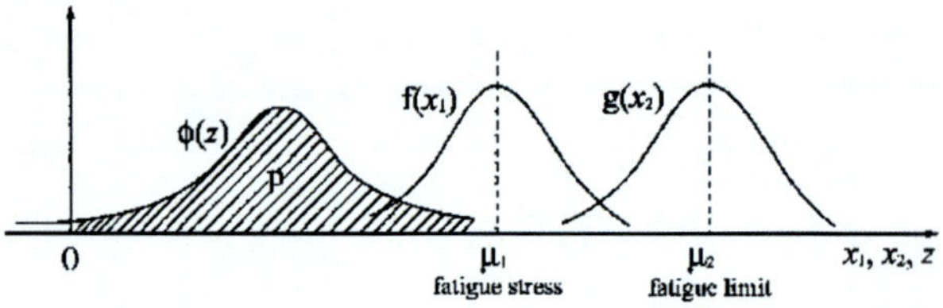

where σ_1 and σ_2 are population standard variations of fatigue stress and limit respectively and assuming that the population mean μ_1 of fatigue stress is less than the population mean μ_2 of fatigue limit. Then structural reliability level p is the probability of x_2 greater than x_1, or:

$$p = P(x_2 > x_1) = P(x_2 - x_1 > 0) \tag{5.3}$$

Letting $z = x_2\_x_1$, then Eq. 5.3 becomes:

$$p = P(z > 0) \tag{5.4}$$

According to statistics principles, it is known that z is also a normal variable with the population mean and standard variation as:

$$\mu = \mu_2 - \mu_1 \tag{5.5}$$

$$\sigma = \sqrt{\sigma_1^2 + \sigma_2^2} \tag{5.6}$$

So the PDF of z is

$$\varphi(z) = \frac{1}{\sqrt{\sigma_2^2 + \sigma_1^2}\sqrt{2\pi}} \exp\left\{ -\frac{[z - (\mu_2 - \mu_1)]^2}{2(\sigma_2^2 + \sigma_1^2)} \right\} \tag{5.7}$$

The PDF curve of z is shown in Fig. 5.1. From Eq. 5.4, it is possible to obtain:

$$p = P(x_2 - x_1 > 0) = \int_0^\infty \varphi(z)dz \tag{5.8}$$

It is clear that p is equal to the dashed area in Fig. 5.1. As shown in Fig. 5.1, $z_p = 0$, from Eq. 2.41, it is possible to have:

$$u_p = \frac{z_p - \mu}{\sigma} = -\frac{\mu}{\sigma} \tag{5.9}$$

and

$$-u_p = \frac{\mu_2 - \mu_1}{\sqrt{\sigma_1^2 + \sigma_2^2}} \tag{5.10}$$

Equation 5.10 is known as the coupling equation to describe a relationship between fatigue stress, fatigue limit and reliability level. In engineering application, the population parameters μ_1, μ_2, σ_1 and σ_2 in Eq. 5.10 are replaced by their estimators with a confidence level of γ. Therefore, in the case where Eq. 5.10 is satisfied, it implies that a fatigue crack will not occur in the structure during an infinitely long service period at a reliability level of p and a confidence level of γ. Note that Eq. 5.10 can be used only when both fatigue stress and strength follow normal distributions. If either of them follows another distribution, then this model is inapplicable for engineering design. Thus a universal formula of one-dimensional stress-strength interference model useful for any distribution needs to be developed.

5.3 A Generalised Interference Model

If the one-dimensional distributions of fatigue stress amplitude s_a and strength amplitude S_a are $f(s_a)$ and $g(S_a)$ respectively (as shown in Fig. 5.2), then for any stress amplitude s_a, the probability of S_a being less than s_a is

$$F(s_a) = P(S_a < s_a) = \int_0^{s_a} g(S_a) dS_a \tag{5.11}$$

Since stress amplitude is a non-negative variable, the lower limit value of integration is zero and the distribution function $F(s_a)$ is shown in dashed area of Fig. 5.2. It is also known that the probability for the occurrence of s_a is $f(s_a)ds_a$. Because $F(s_a)$ is a monotonic function with respect to s_a, and is also a random variable, it is possible to obtain the probability for the occurrence of $F(s_a)$ also being $f(s_a)ds_a$. The arithmetic product of these two entities becomes the differential coefficient df of structural rate-of-failure f as:

$$df = F(s_a)f(s_a)ds_a \tag{5.12}$$

Integrating Eq. 5.12, it is possible to have the structural rate-of-failure f as:

$$f = \int_0^{(s_a)_{max}} F(s_a)f(s_a)ds_a \tag{5.13}$$

So structural reliability level p is:

$$p = 1 - f = 1 - \int_0^{(s_a)_{max}} F(s_a)f(s_a)ds_a \tag{5.14}$$

Substituting Eq. 5.11 into Eq. 5.14, one has:

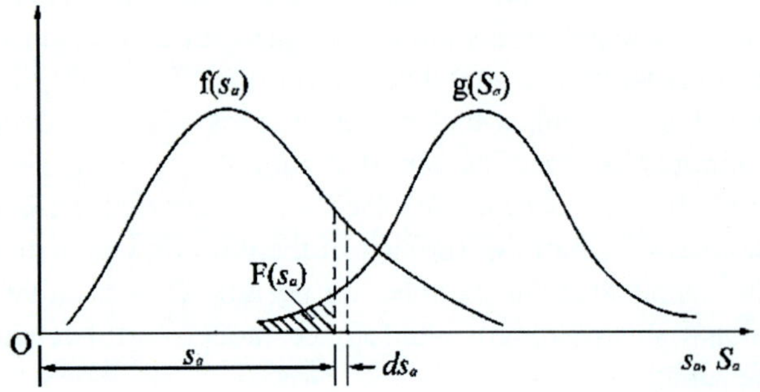

Fig. 5.2 One-dimensional stress-strength interference model

$$p = 1 - \int_{0}^{(s_a)_{max}} \left[\int_{0}^{s_a} g(S_a) dS_a \right] f(s_a) ds_a \tag{5.15}$$

Equation 5.15 is known as the universal formula of one-dimensional stress-strength interference model. As already mentioned, it is well known that fatigue stress and strength are dominated by two-dimensional random variables of (s_a, s_m) and (S_a, S_m) respectively. If their p.d.fs are respectively $f(s_a, s_m)$ and $g(S_a, S_m)$, then using the analogy of Eq. 5.11, the probability of (S_a, S_m) being less than (s_a, s_m) is:

$$F(s_a, s_m) = P[(S_a, S_m) < (s_a, s_m)]$$

$$F(s_a, s_m) = \int_{(s_m)_{min}}^{s_m} \int_{0}^{s_a} g(S_a, S_m) dS_a dS_m \tag{5.16}$$

Similarly, it is known that all probabilities for the occurrences of (s_a, s_m) and $F(s_a, s_m)$ are equal to $f(s_a, s_m)ds_a ds_m$. The arithmetic product between $F(s_a, s_m)$ and its probability of occurrence is the differential coefficient df of structural rate-of-failure f as:

$$df = F(s_a, s_m) f(s_a, s_m) ds_a ds_m \tag{5.17}$$

Integrating Eq. 5.17, it is possible to have the structural rate-of-failure f as:

$$f = \int_{(s_m)_{min}}^{(s_m)_{max}} \int_{0}^{(s_a)_{max}} F(s_a, s_m) f(s_a, s_m) ds_a ds_m \tag{5.18}$$

So the structural reliability level p becomes:

$$p = 1 - f = 1 - \int_{(s_m)_{min}}^{(s_m)_{max}} \int_{0}^{(s_a)_{max}} F(s_a, s_m) f(s_a, s_m) ds_a ds_m \tag{5.19}$$

Substituting Eq. 5.16 into Eq. 5.19, the two-dimensional stress-strength interference model can be obtained as:

$$p = 1 - \int_{(s_m)_{min}}^{(s_m)_{max}} \int_{0}^{(s_a)_{max}} \left[\int_{(s_m)_{min}}^{s_m} \int_{0}^{s_a} g(S_a, S_m) dS_a dS_m \right] f(s_a, s_m) ds_a ds_m \tag{5.20}$$

If an adequately high reliability level such as $p = 99.99\%$ is given, then the probability of fatigue occurrence will reduce significantly.

5.4 Fracture Interference Model

The interference model mentioned above can be also applied for infinite life design of notched structures, implying that an initial crack, if it exists will not propagate during an infinitely long service period at a reliability level of p and a confidence level of γ. In this case, μ_1 and σ_1 are population mean and standard variation of the stress intensity factor range ΔK. μ_2 and σ_2 are population mean and standard variation of facture threshold ΔK_{th}. For a safe operation assurance of a notched structure subjected to steady stochastic cyclic stress, the stress intensity factor range should be controlled to be under the facture threshold. Therefore, according to the analogy of stress-strength interference model, a fracture interference model needs to be established for damage tolerance design of infinite life and fault analysis.

By means of Eqs. 1.48 and 1.49, the stress intensity factor mean K_m and range ΔK can be written as:

$$K_m = X(s_{\text{m}}) \cdot Y(a) \tag{5.21}$$

$$\Delta K = [X(s_{\max}) - X(s_{\min})] \cdot Y(a) = \Delta X \cdot Y(a) \tag{5.22}$$

From Eqs. 5.21 and 5.22, it is seen that the probability distributions of the stress intensity factor mean K_m and range ΔK at a given crack length a can be derived from the probability distributions of the stress level. Assuming that one-dimensional probability density functions of the stress intensity factor range ΔK and facture threshold ΔK_{th} are $f(\Delta K)$ and $g(\Delta K_{\text{th}})$ respectively, then for any of ΔK, the probability of ΔK_{th} being less than ΔK is:

$$F(\Delta K) = P(\Delta K_{\text{th}} < \Delta K) = \int_0^{\Delta K} g(\Delta K_{\text{th}}) d\Delta K_{\text{th}} \tag{5.23}$$

Because $F(\Delta K)$ is a monotonic function with respect to the random variable ΔK, the probability for the occurrence of $F(\Delta K)$ is $f(\Delta K)d\Delta K$, and the differential coefficient df of structural rate-of-failure f is:

$$df = F(\Delta K) f(\Delta K) d\Delta K \tag{5.24}$$

Integrating Eq. 5.24 leads to the structural rate-of-failure f:

$$f = \int_0^{(\Delta K)_{\max}} F(\Delta K) f(\Delta K) d\Delta K \tag{5.25}$$

So the probability of a crack not propagating, or structural reliability level p is:

$$p = 1 - f = 1 - \int_0^{(\Delta K)_{\max}} F(\Delta K) f(\Delta K) d\Delta K \tag{5.26}$$

Substituting Eq. 5.23 into Eq. 5.26, it is possible to have:

$$p = 1 - \int_0^{(\Delta K)_{\max}} \left[\int_0^{\Delta K} g(\Delta K_{\text{th}}) d\Delta K_{\text{th}} \right] f(\Delta K) d\Delta K \tag{5.27}$$

Equation 5.27 is known as the one-dimensional fracture interference model. In the same manner, Eq. 5.27 can be extended to a two-dimensional situation. If the probability density functions of the stress intensity factor range and mean, and the facture threshold range and mean are $f(\Delta K, K_m)$ and $g(\Delta K_{\text{th}}, K_{m\text{th}})$ respectively, then the two-dimensional fracture interference model can be written as:

$$p = 1 - \int_{(K_m)_{\min}}^{(K_m)_{\max}} \int_0^{(\Delta K)_{\max}} \left[\int_{(K_m)_{\min}}^{K_m} \int_0^{\Delta K} g(\Delta K_{\text{th}}, K_{m\text{th}}) d\Delta K_{\text{th}} dK_{m\text{th}} \right] \times f(\Delta K, K_m) d\Delta K dK_m \tag{5.28}$$

For a given initial crack size a_0, from Eq. 5.27 or 5.28, the reliability level p relating to non-propagation of a crack pertinent to this crack size can be predicted. Equally, for a given reliability level of p, the maximum crack size a_{cr} for non-propagation of the crack pertaining to this reliability level can also be determined by an inverse procedure based on Eq. 5.27 or 5.28.

5.5 Reduction Factor

The structural safe life of an engineering artefact with a high probability of survival and with a high confidence level is determined from the complete set of fatigue test data by means of a scatter factor. Further, a reduction factor for fatigue strength is used for high cycle fatigue reliability design. The reduction factor r of fatigue strength is defined as the ratio of safe fatigue strength $\hat{S}_p$ to the sample median fatigue strength $[S_{50}]$, i.e.,

$$r = \frac{\hat{S}_p}{[S_{50}]} \tag{5.29}$$

where $[S_{50}]$ represents the estimator of sample median fatigue strength with 50% reliability level.

In case that fatigue strength follows a log normal distribution with the mean of μ and standard deviation of σ (or $x = \log S$ follows $N(\mu, \sigma)$), the estimator $\hat{x}_p$ of logarithmic safe fatigue strength pertinent to reliability level p can be written as:

$$x_p = \mu + u_p \sigma$$

If the population standard deviation is known, i.e., $\sigma = \sigma_0$, the estimator $\hat{x}_p$ of logarithmic safe fatigue strength $x_p = \mu + u_p\sigma$ pertinent to reliability level p becomes:

$$\hat{x}_p = \hat{\mu} + u_p\sigma_0 \tag{5.30}$$

where u_p is the standard normal deviator pertinent to the reliability level of p; $\hat{\mu}$ is the estimator of population mean μ and is generally replaced with the sample mean $\bar{x}$. Note that the sample mean $\hat{\mu}$ is different from the population mean μ. Thus, a positive constant C is introduced so as to prevent the predicted value of $\hat{\mu}$ for the sample from inclining to the low side. $\hat{\mu}$ can be written as:

$$\hat{\mu} = \bar{x} - C \tag{5.31}$$

The present question is to find the value of C in such a way that with a confidence level of γ, the random variable $\hat{\mu}$ is less than the true μ, i.e.,

$$P(\hat{\mu} = \bar{x} - C > \mu) = \gamma \tag{5.32}$$

where γ is the confidence level. Equation 5.31 states that the probability of the random variable $(\bar{X} - C)$ being greater than μ is γ, or, the estimator $\hat{\mu}$ is greater than μ at a given confidence level γ.

From statistics, it is well known that if $\bar{X}$ follows a Gauss distribution of $N(\mu, \sigma_0/\sqrt{n})$ then $(\bar{X} - C)$ also follows a Gauss distribution of $N(\mu - C, \sigma/\sqrt{n})$. The p.d.f of random variable $\bar{X} - C$ is shown in Fig. 5.3. From Fig. 5.3 and Eq. 5.31, the relationship between μ and $(\mu - C)$ can be obtained as:

$$\mu = \mu - C + u_\gamma\frac{\sigma_0}{\sqrt{n}} \tag{5.33}$$

where u_γ is the standard normal deviator pertinent to the confidence level of γ. From Eq. 5.33, one can get:

$$C = u_\gamma\sigma/\sqrt{n} \tag{5.34}$$

Substituting Eq. 5.34 into Eq. 5.31 yields the estimator of population mean as:

$$\hat{\mu} = \bar{x} - u_\gamma\frac{\sigma_0}{\sqrt{n}} \tag{5.35}$$

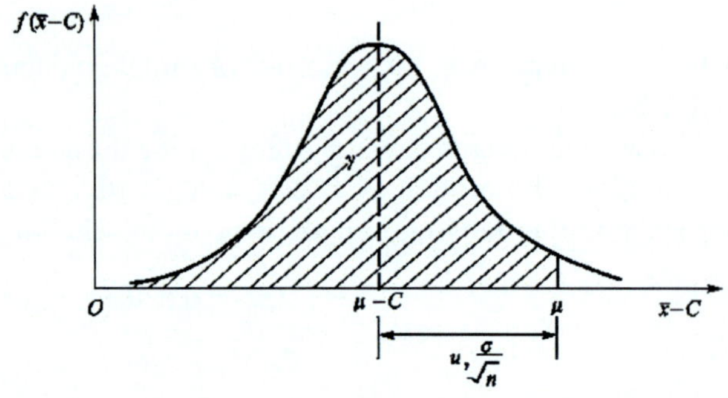

Fig. 5.3 PDF of random variable $\bar{X} - C$

Again, substituting Eq. 5.35 into Eq. 5.30 gives:

$$\hat{x}_p = \bar{x} - u_\gamma \frac{\sigma_0}{\sqrt{n}} + u_p \sigma_0 \tag{5.36}$$

Equation 5.36 can be written as:

$$\log \hat{S}_p = \log[S_{50}] - u_\gamma \frac{\sigma_0}{\sqrt{n}} + u_p \sigma_0 \tag{5.37}$$

Transformation of Eq. 5.37 gives:

$$\log \frac{\hat{S}_p}{[S_{50}]} = \left(u_p - \frac{u_\gamma}{\sqrt{n}} \right) \sigma_0 \tag{5.38}$$

From the definition formula (5.29) of reduction factor, it is possible to have:

$$r = 10^{\left(u_p - \frac{u_\gamma}{\sqrt{n}} \right) \sigma_0} \tag{5.39}$$

where n is the number of specimen; σ_0 is the population standard deviation and can be calculated from the existing larger sample numbers of experimental data.

In case of small sample numbers, there is an error between the predictions $\bar{x}$ and $\hat{\sigma}$ for sample mean and standard deviation of normal random variable X and the population true values μ and σ. According to statistics, the predictions $\hat{x}_p$ for sample percentile of normal random variable X can be obtained:

$$\hat{x}_p = \bar{x} + u_p \hat{\sigma} \tag{5.40}$$

Obviously, there is an error between the predictions $\hat{x}_p$ for sample percentile of normal random variable X, determined from Eq. 5.40, and the population true value $x_p = \mu + u_p \sigma$. As a result, the confidence level of statistical results estimated from small samples need to be analysed. From Eq. 2.95, the estimator $\hat{x}_p$ of logarithmic safe fatigue strength $x_p = \mu + u_p \sigma$ corresponding to the probability of survival of p and the confidence level of γ is:

$$\hat{x}_p = \left(\bar{x} + u_p \hat{k} s \right) - t_\gamma s \sqrt{\frac{1}{n} + u_p^2 \left(\hat{k}^2 - 1 \right)} \tag{5.41}$$

Transforming Eq. 5.41 shows:

$$\log \frac{\hat{S}_p}{[S_{50}]} = \left(\bar{x} + u_p \hat{k} s \right) - t_\gamma s \sqrt{\frac{1}{n} + u_p^2 \left(\hat{k}^2 - 1 \right)}$$

Thus:

$$r = \frac{\hat{S}_p}{[S_{50}]} = 10^{u_p \hat{k} s - t_\gamma s \sqrt{1/n + u_p^2 (\hat{k}^2 - 1)}} o \tag{5.42}$$

In case fatigue strength follows a normal distribution with a mean of μ and standard deviation of σ_0, and the population standard deviation is known, or $\sigma = \sigma_0$, then from Eq. 5.36, the estimator $\hat{x}_p$ of safe fatigue strength $x_p = \mu + u_p\sigma$ pertinent to reliability level p can be written as:

$$\hat{x}_p = \bar{x} - u_\gamma \frac{\sigma_0}{\sqrt{n}} + u_p\sigma_0 \tag{5.43}$$

Or:

$$\hat{S}_p = [S_{50}] - u_\gamma \frac{\sigma_0}{\sqrt{n}} + u_p\sigma_0 \tag{5.44}$$

Through transformation, Eq. 5.44 becomes:

$$\frac{\hat{S}_p}{[S_{50}]} = 1 + \left(u_p - \frac{u_\gamma}{\sqrt{n}}\right)\frac{\sigma_0}{[S_{50}]} \tag{5.45}$$

Based on the definition of coefficient of variability, it is possible to have:

$$C_v = \frac{\sigma_0}{[S_{50}]} \tag{5.46}$$

From Eq. 5.29, substituting Eq. 5.46 into Eq. 5.45 shows:

$$r = 1 + C_v\left(u_p - \frac{u_\gamma}{\sqrt{n}}\right) \tag{5.47}$$

In case of small sample numbers, i.e., the population standard deviation is unknown, using the analogy of Eq. 5.42, it can be shown that:

$$\frac{\hat{S}_p}{[S_{50}]} = 1 + u_p C_v - t_\gamma C_v \sqrt{\frac{1}{n\hat{k}^2} + u_p^2\left(1 - \frac{1}{\hat{k}^2}\right)}$$

Thus:

$$r = 1 + u_p\hat{k}C_v - t_\gamma C_v\sqrt{\frac{1}{n} + u_p^2(\hat{k}^2 - 1)} \tag{5.48}$$

5.6 Scatter Factor

The safe structural life of an engineering artefact with a high probability of survival and with a high confidence level is determined from a complete set of fatigue test data by means of the scatter factor method. The scatter factor is the ratio between the best estimator of the fatigue life obtained from a small number of tests and the time to the first failure in an actual structure at a specified reliability level. The scatter factor S_f for fatigue crack initiation defined by median life is:

$$S_f = \frac{N_{50}}{N_p} \tag{5.49}$$

Usually, N_{50} is replaced by the sample median life $\hat{N}_{50}$; then the scatter factor S_f for fatigue crack initiation is defined by sample median life as:

$$S_f = \frac{\hat{N}_{50}}{\hat{N}_p} \tag{5.50a}$$

Alternatively, it can be written as:

$$S_f = \frac{\hat{N}_{50}}{\hat{N}_p} = 10^{-\mu_p \sigma} \tag{5.50b}$$

where p is the reliability level, σ is the population standard deviation of crack initiation life, $\hat{N}_{50}$ is the sample median crack initiation life, $\hat{N}_p$ is the safe fatigue crack initiation life, or probable number of cycles at which visually detectable cracks are formed and u_p is the standard normal deviator corresponding to the reliability level of p.

By analogy of Eq. 5.50b, it is possible to have the definition of the scatter factor for fatigue crack propagation as:

$$S_f = \frac{\hat{N}_{50}^*}{\hat{N}_p^*} = 10^{-u_p \sigma^*} \tag{5.51}$$

where σ^* is the population standard deviation of crack propagation life, $\hat{N}_{50}^*$ is the sample median crack propagation life and $\hat{N}_p^*$ is the probable fatigue crack propagation or probable number of cycles at which the critical crack length is reached, i.e., failure is imminent.

Equation 5.50b assumes a log-normal distribution for fatigue life N. Letting $Y = \log N$, then the random variable Y follows a normal distribution and the population percentile y_p of normal random variable Y pertinent to reliability level p is:

$$y_p = \mu + u_p \sigma \tag{5.52}$$

Or:

$$N_p = 10^{(\mu + u_p \sigma)} \tag{5.53}$$

where μ is the population mean value and σ is the population standard deviation. The sample percentile $\hat{y}_p$ of normal random variable Y pertinent to reliability level p is:

$$\hat{y}_p = \hat{\mu} + u_p \sigma = \log \hat{N}_{50} + u_p \sigma \tag{5.54}$$

Or:

$$\hat{N}_p = \hat{N}_{50} \times 10^{(u_p \sigma)} \tag{5.55}$$

In case the population standard deviation is known, or $\sigma = \sigma_0$, then from Eq. 5.36, the estimator $\hat{y}_p$ of safe logarithmic fatigue life $y_p = \mu + u_p\sigma$ pertinent to reliability level p can be written as:

$$\hat{y}_p = \bar{y} - u_\gamma \frac{\sigma_0}{\sqrt{n}} + u_p \sigma_0 \tag{5.56}$$

Or:

$$\log \hat{N}_p = \log \hat{N}_{50} - u_\gamma \frac{\sigma_0}{\sqrt{n}} + u_p \sigma_0 \tag{5.57}$$

Through transforming Eq. 5.57, it is possible to have:

$$\log \frac{\hat{N}_{50}}{\hat{N}_p} = \left(\frac{u_\gamma}{\sqrt{n}} - u_p \right) \sigma_0 \tag{5.58}$$

From Eqs. 5.50a and 5.58, one has:

$$S_f = 10^{\left(\frac{u_\gamma}{\sqrt{n}} - u_p \right) \sigma_0} \tag{5.59}$$

where n is the sample size, γ is the confidence level, σ_0 is the population standard deviation of logarithmic fatigue life and u_γ is the standard normal deviator corresponding to the confidence level of γ.

By analogy of Eq. 5.59, it is possible to have the scatter factor for fatigue crack propagation as:

$$S_f = 10^{\left(\frac{u_\gamma}{\sqrt{n}} - u_p \right) \sigma_0^*} \tag{5.60}$$

For different types of metallic materials, the population standard deviation σ_0 of logarithmic fatigue crack initiation life has been recommended to be $\sigma_0 = 0.16 \sim 0.20$, while the population standard deviation σ_0^* of logarithmic fatigue crack propagation life has been suggested to be $\sigma_0^* = 0.07 \sim 0.10$ [14]. It is worth pointing out that for welded steel details, the population standard deviations could be larger than those indicated above [15]. Equation 5.59 has been widely used, but its application conditions are: (a) the fatigue test results of full-scale structure must be complete data; and (b) the population standard deviation is given, $\sigma = \sigma_0$. When these conditions are not met, Eq. 5.59 is inapplicable.

If the population standard deviation σ_0 is unknown, from Eq. 2.95, the estimated value of logarithmic safe life can be obtained as:

$$\log \hat{N}_p = \left(\log \hat{N}_{50} + u_p \hat{k} s \right) - t_\gamma s \sqrt{\frac{1}{n} + u_p^2 \left(\hat{k}^2 - 1 \right)} \tag{5.61}$$

Thus, it is possible to determine the service life approximately using the following scatter factor:

$$S_f = 10^{u_p \hat{k} s - t_\gamma s \sqrt{1/n + u_p^2 (\hat{k}^2 - 1)}} \tag{5.62}$$

where s is the sample standard deviation and t_γ is the γ percentile of t-distribution, β is the correction coefficient of the standard deviation. It is worth noting that the standard deviation is a biased statistic, in that there is a difference between the sample and population standard deviations. In case of small sample sizes, the difference is great and the correction coefficient needs to correct the sample standard deviation.

Moreover, it is often the case that full-scale fatigue test results are incomplete because of cost and time constraints. Thus it is desirable to have a technique that accounts for the incompleteness of the dataset. Though the Maximum Likelihood Method is a rational and easy way in the statistical evaluation of non-failed fatigue tests [16], it still is expedient and effective for the rank technique to statistically analyse small sample of incomplete data. Consider, as an example, the case of the nominally identical port and starboard wings of an aeroplane undergoing fatigue testing. If either of the two parts fails under a given load spectrum, the testing stops. The service life of the structure needs to be determined from this incomplete dataset. Let $x_1 = \log N_1$, and $x_2 = \log N_2$, where N_1 is the fatigue life of a broken structure, and N_2 is the fatigue life of an unbroken structure. It is obvious that $x_2 > x_1$, and the estimation formula of the probability of survival is:

$$\hat{p} = 1 - \frac{i}{n+1} \tag{5.63}$$

From the formula above, the estimated values of probability of survival corresponding to x_1 and x_2 are respectively:

$$\hat{p}_1 = 1 - \frac{1}{2+1} = \frac{2}{3}$$

and

$$\hat{p}_2 = 1 - \frac{2}{2+1} = \frac{1}{3}$$

As both of x_1 and x_2 come from the same normal population $N(\mu, \sigma_0)$, we have:

$$x_1 = \mu + u_1 \sigma_0 \tag{5.64a}$$

$$x_2 = \mu + u_2 \sigma_0 \tag{5.64b}$$

where u_1 and u_2 are the standard normal deviation variables. Transformation of Eq. 5.64 gives:

$$x_2 = x_1 + (u_2 - u_1)\sigma_0 \tag{5.65}$$

Then:

$$\log \hat{N}_{50} = \frac{1}{2}(x_1 + x_2) = x_1 + \frac{1}{2}(u_2 - u_1)\sigma_0 \tag{5.66}$$

Substituting Eq. 5.50a into formula (5.59), the service life with reliability level of p can be calculated according to the following equation:

$$\hat{N}_p = \frac{\hat{N}_{50}}{S_f} = 10^{\left[x_1 + \frac{1}{2}(u_2 - u_1)\sigma_0 + u_p\sigma_0 - \frac{u_\gamma}{\sqrt{n}}\sigma_0\right]} \tag{5.67}$$

Thus:

$$\log \hat{N}_p = \log N_1 + \frac{1}{2}(u_2 - u_1)\sigma_0 + u_p\sigma_0 - \frac{u_\gamma}{\sqrt{n}}\sigma_0 \tag{5.68}$$

By means of Eqs. 5.66 and 5.68, the scatter factor for the incomplete test data can be derived as follows:

$$S_f = \frac{N_1}{\hat{N}_p} = 10^{\left(\frac{u_\gamma}{\sqrt{2}} - u_p\right)\sigma_0 - \frac{1}{2}(u_2 - u_1)\sigma_0} \tag{5.69}$$

From the normal distribution function tables [13] it is possible to obtain the standard normal deviation variable $u_1 = -0.4308$ pertaining to a reliability level of $p = 2/3$ and $u_2 = 0.4308$ pertinent to $p = 1/3$, and so

$$S_f = 10^{\left(\frac{u_\gamma}{\sqrt{2}} - u_p - 0.4308\right)\sigma_0} \tag{5.70}$$

5.7 Durability Model to Assess Economic Structural Life

In the last two decades, theory and methods of structural reliability assessment have been developed significantly to achieve the inspection and maintenance plans for ensuring structural integrity at the design stage and pure failure probability technique [17, 18] has been applied for reliability formulation to estimate the failure probabilities of the components and the operational safety is ensured by regular inspections in combination with mechanical calculations. Such plans can be expensive. In recent years, owing to the need to reflect an economic point of view, risk-based inspection methodologies [19, 20] are starting to become prevalent in order to consider the consequences (cost) of failure so as to minimise costs but satisfying a minimum reliability level. Here risk is generally defined as the consequence (cost) multiplied by the probability of failure to address the comprehensive effect of structural safety and economic requirement.

Because the fatigue process includes the two-stages of crack initiation and crack propagation, the total fatigue life of a structural component is then equal to the sum of the time to crack initiation (i.e., to visually detectable crack size a_0) and the time for the crack to reach its critical size a_{cr}. Therefore it is necessary for structural life to be assessed using a combination of safe life methodology and damage tolerance approach. The safe life method is applied for life assessment and damage tolerance approach is for operational assessment of safety. Based on this philosophy, a life assessment model can be developed for a finite life design [21]. An approach could be based using probable lives at damage inception and when cracks reach critical dimensions. So the first inspection and maintenance period T_1 should satisfy the condition of $T = T^*$: the condition $T < T^*$ implies that the structure does not need to be inspected as yet.

For a large or whole structure comprising several components, if failure of any component will result in the failure of whole structure, the structure can be then regarded as a series system with $i = 1, 2, \ldots, m$ series elements (components). If the reliability level of the ith component is denoted as R_i, then the reliability of whole structure becomes:

$$R = \prod_{i=1}^{m} R_i \tag{5.71}$$

Equation 5.71 shows that the independence between elements is assumed to idealise the reliability formulation from an engineering approximation point of view, the latter reflecting common practice that failure of any component is unacceptable.

Assuming A, B and C denote respectively the events of crack initiation, crack propagation until its critical size a_{cr}, and no initial crack detection in a structural component, then $(C \cap A \cap B)$ and $(\bar{C} \cap B)$ represent the events of no initial crack detection and of an initial crack detection and propagation until critical size a_{cr} respectively. Then the failure probability of the structural component is:

$$F = P\{(C \cap A \cap B) \cup (\bar{C} \cap B)\} \tag{5.72}$$

From probability theory, expanding Eq. 5.72 gives:

$$F = P(C \cap A \cap B) + P(\bar{C} \cap B) = P(C)P(A \cap B|C) + P(\bar{C})P(B|\bar{C}) \tag{5.73}$$

From this it is possible to show:

$$F = P(C)P(A|C)P(B|A \cap C) + P(\bar{C})P(B|\bar{C}) \tag{5.74}$$

where:

$$P(C) = p_0 \tag{5.75}$$

$$P(\bar{C}) = 1 - p_0 \tag{5.76}$$

$$P(B|\bar{C}) = 1 - p^* \tag{5.77}$$

$$P(A|C) = 1 - p \tag{5.78}$$

$$P(B|A \cap C) = 1 - p^* \tag{5.79}$$

Here $(1-p)$ represents the probability of crack initiation. $(1-p^*)$ implies the probability of crack grows until the critical crack. P(C) is dependent on the crack size and is the probability p_0 of no initial crack detection in a structural component when first taken into service. In general, the size of a detected crack is measured by a non-destructive method for structures in service and each detection technique has a limiting size of detection, a_0. Because the real measurements of a crack size usually involve a large scatter on the probability of detection, e.g., measuring inaccuracies, therefore the inspection capability may be described by the probability of detection [22]:

$$p_0 = \begin{cases} \exp\left[-\frac{2a-a_0}{\lambda_0}\right], & 2a > a_0 \\ 1, & 2a \leq a_0 \end{cases} \tag{5.80}$$

where λ_0 is the characterised parameter which has values between 0 and ∞. a_0 is the limit size of detection.

Substituting Eqs. 5.75–5.79 into Eq. 5.71 yields:

$$F = p_0(1-p)(1-p^*) + (1-p_0)(1-p^*) \tag{5.81}$$

From Eq. 5.81, it is possible to have the reliability level of the ith component:

$$R_i = 1 - p_{0i}(1-p_i)\left(1-p_i^*\right) - (1-p_{0i})\left(1-p_i^*\right) \tag{5.82}$$

Again, substituting Eq. 5.82 into Eq. 5.71 gives:

$$R = \prod_{i=1}^{m}\left[1 - p_{0i}(1-p_i)\left(1-p_i^*\right) - (1-p_{0i})\left(1-p_i^*\right)\right] \tag{5.83}$$

The reliability levels p and p^* of possible fatigue crack initiation and propagation in a structural component pertaining to the crack initiation and propagation lives of T and T^* can be determined from a small number of experimental results or calculations by means of the Eqs. 1.18, 1.54, 1.58 and 1.62 using scatter factor formulations (5.50) and (5.51). p and p^* can be written as:

$$p = f(T) \tag{5.84a}$$

$$p^* = g(T^*) \tag{5.84b}$$

and substituting Eqs. 5.84a and 5.84b into Eq. 5.83 results in:

$$R = \prod_{i=1}^{m}\left\{1 - p_{0i}[1 - f_i(T_i)]\left[1 - g_i^*\left(T_i^*\right)\right] - (1-p_{0i})\left[1 - g_i^*\left(T_{0i}^*\right)\right]\right\} \tag{5.85}$$

It is worth noting that Eq. 5.85 have described possible cases of initial crack detection, crack initiation and crack propagation, but have not considered the case of crack repair and the effect of more than one inspection (in time or in different components) yet. However, it is apparent that Eq. 5.85 can be used to realistically and effectively to determine the first inspection and maintenance period of a new structure when first used in service in an easy way.

After the first inspection and maintenance, by analogy of Eqs. 5.84a and 5.84b, one has the new p–T and p^*–T^* curves of a structural component. Using the analogy of Eq. 5.85 again, it is possible to have a new reliability formulation to determine the next inspection and maintenance period for the structure.

In order to obtain the first inspection and maintenance period of a structure when first used, it is possible to assume that the structure is new and there is no visibly detectable crack. This implies that the probability p_0 of no initial crack detection in a new structure when first used is 1. Thus, Eq. 5.85 becomes:

$$R = \prod_{i=1}^{m} \left[f_i(T_i) + g_i\left(T_i^*\right) - f_i(T_i) \cdot g_i\left(T_i^*\right)\right] \tag{5.86}$$

From Eq. 5.86, it is possible to obtain the reliability of only a structural component:

$$R = f(T_p) + g\left(T_p^*\right) - f(T_p) \cdot g\left(T_p^*\right) \tag{5.87}$$

From Eq. 5.87 and the condition of $T = T^*$ satisfied by the first inspection and maintenance period, T_1 can be determined by the following equation:

$$R = f(T_1) + g(T_1) - f(T_1) \cdot g(T_1) \tag{5.88}$$

Equation 5.88 represents the outcome of the durability model to determine the first inspection and maintenance period. It is worth pointing out that Eqs. 5.85–5.88 are applied for only determining the inspection and maintenance period of a new structure when first used.

Example 5.1 A fastening structure at the root of helicopter blade consists of the auricle junction, bolts and the blade beam: the auricle junction and bolts are made of 40CrNiMoA alloyed steel while the blade beam is made of LD_2 aluminium alloy. The geometry and dimensions of fastening structure are illustrated in Fig. 5.4. From engineering practice, it is clear that first fatigue failure of fastening structure appears on the auricle junction; it is therefore important to assess the economic first inspection and maintenance period of the auricle junction. The mechanical properties of 40CrNiMoA alloyed steel are as follows: Young's modulus $E = 204$ GPa, Poisson's ratio $v = 0.3$, ultimate strength $\sigma_b = 1080$ MPa, yield stress $\sigma_{0.2} = 880$ MPa, fracture toughness $K_{IC} = 4691 \quad \text{MPa}\sqrt{\text{mm}}$, fracture threshold value $\Delta K_{th} = 342 \quad \text{MPa}\sqrt{\text{mm}}$, mean fatigue limit $S_0 = 75.4$ MPa and both population standard deviations of logarithmic fatigue crack initiation and propagation life $\sigma_0 = \sigma_0^* = 0.2$ [23]. From Eqs. 1.18 and 1.44, the S_a–S_m–N surface and *da/dN*–ΔK curve of the auricle junction are respectively given as:

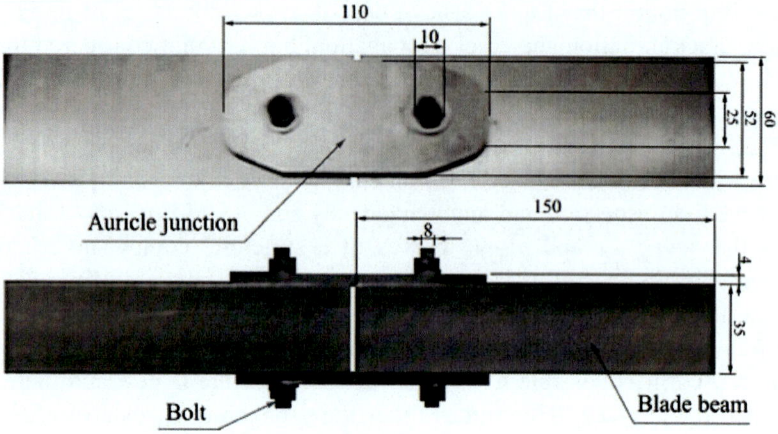

Fig. 5.4 Fastening structure at the root of helicopter blade (all dimension in mm)

$$N = 33497 \times \left(\frac{14.32}{1080.0 - S_m} S_a - 1 \right)^{-1.2402} \tag{5.89}$$

$$\frac{da}{dN} = 3.62 \times 10^{-9.0} \times \left[\left(\frac{1-f}{1-r} \right) \Delta K \right]^{1.715} \cdot \frac{\left(1 - \frac{342.0}{\Delta K}\right)^{2.314}}{\left[1 - \frac{\Delta K}{4691.0 \times (1-r)}\right]^{0.285}} \tag{5.90}$$

The nominal stress spectrum (shown in Fig. 5.5) can be obtained from the actual load spectrum. A block of the nominal stress spectrum represents the actual load history of one flight hours. Consequently, by means of Eq. 1.58, the fatigue crack initiation life can be calculated from the nominal stress spectrum (shown in Fig. 5.5) and the S_a–S_m–N surface formulation (5.89) of the auricle junction. The safe fatigue crack initiation lives with a series of reliability level are calculated by using Eq. 5.50b. The calculations are shown in Table 5.1.

Fatigue crack growth mode of the auricle junction can be simulated through a single side-penetrated crack along the notched edge of a finite-width plate subjected a cyclic from the bolt (shown in Fig. 5.6). The geometry function of fatigue crack of auricle junction is then determined as:

$$\alpha(a) = (0.707 - 0.18\lambda + 6.55\lambda^2 - 10.54\lambda^3 + 6.85\lambda^4) \cdot \left[\frac{\pi\varphi}{2W} + \frac{1}{\pi}\left(\frac{\varphi}{\varphi + a} \right)\sqrt{\lambda} \right] \cdot \sqrt{\sec\left(\frac{\pi}{2} \cdot \frac{\varphi + a}{W - a} \right) \cdot \sec\left(\frac{\pi\varphi}{2W} \right)} \tag{5.91}$$

with

$$\lambda = \frac{\varphi}{\varphi + 2a}$$

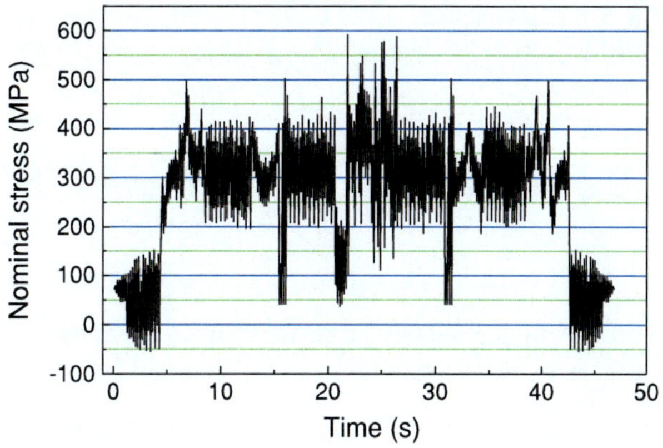

Fig. 5.5 Nominal stress spectrum

Table 5.1 The calculations of safe fatigue crack initiation life

Reliability level p	Safe crack initiation life T_p (flight hour)
0.9999990	1129
0.9999981	1300
0.9999971	1433
0.9999961	1531
0.9999951	1618
0.9999940	1691
0.9999932	1749
0.9999922	1809
0.9999911	1871
0.9999900	1925
0.9999793	2297
0.9999709	2548
0.9999609	2762
0.9999519	2927
0.9999409	3100
0.9999305	3247
0.9999216	3360
0.9999116	3478
0.9999000	3603
0.9997999	4189
0.9996982	4526
0.9995959	4786
0.9994991	4990
0.9993590	5200
0.9992886	5344
0.9991836	5492

The critical crack length a_{cr} of the auricle junction is calculated to be 12.0 mm from fracture toughness K_{1C}. The initial crack size is chosen to be 1.25 mm, by virtue of being detectable through visual inspections. Substituting Eq. 5.91 into

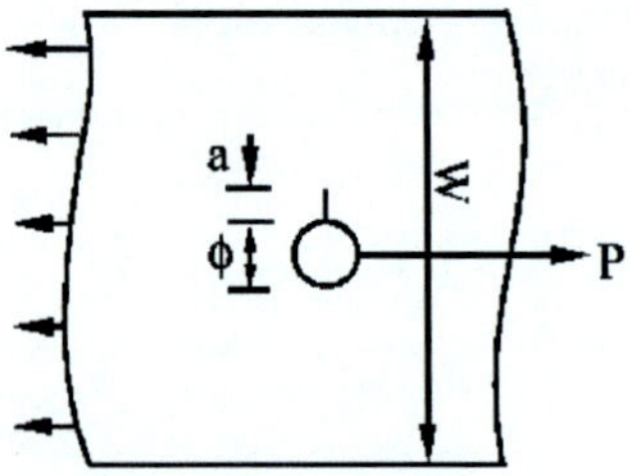

Fig. 5.6 Crack growth model for auricle junction

5.90 and taking integral transformation, the fatigue crack propagation S_a–S_m–N surfaces of auricle junction can be obtained as:

$$N = \frac{10^{9.0} \cdot (2s_a)^{2.314}}{3.62 \cdot [(1-f)(s_a + s_m)]^{1.715}} \int_{1.25}^{12.0} \frac{[2s_a Y(a) - \Delta K_{th}]^{2.314}}{[Y(a)]^{0.599}\left[1 - \frac{s_a + s_m}{K_{1C}} Y(a)\right]^{0.285}} da \quad (5.92)$$

As a result, using Eq. 1.62, the fatigue crack propagation life can be determined from the nominal stress spectrum (shown in Fig. 5.5) and the s_a–s_m–N curve Eq. 5.92 of the auricle junction. Safe fatigue crack propagation lives with a series of reliability levels are determined based on Eq. 5.51. The calculations are shown in Table 5.2.

For a high specific reliability level of 0.999999 to ensure the safety of auricle junction in the lifetime, or from Eq. 5.72, one has:

$$0.999999 = 1 - (1-p)(1-p^*) \quad (5.93)$$

Based on Eq. 5.93, for the required reliability level of 0.999999, a series of solution set of the reliability levels p and p^* of possible crack initiation and propagation can be determined. According to Eqs. 5.84a and 5.84b, it is possible to have a series of solution sets of the safe fatigue crack initiation and propagation lives (denoted respectively by T and T^*) corresponding to the reliability levels of p and p^*. In the plane coordinate system of T–T^*, a set of T and T^* can be displayed as a point and a series of solution sets of T and T^* constitute a T–T^* curve. Therefore, it is obvious that there are a series of solution sets of T and T^*

Table 5.2 The calculations of safe fatigue crack propagation life

Reliability level p^*	Safe crack propagation life T_p^* (flight hour)
0.999999	44
0.999990	55
0.999900	71
0.999000	95
0.998775	98
0.997525	108
0.995002	120
0.990000	135

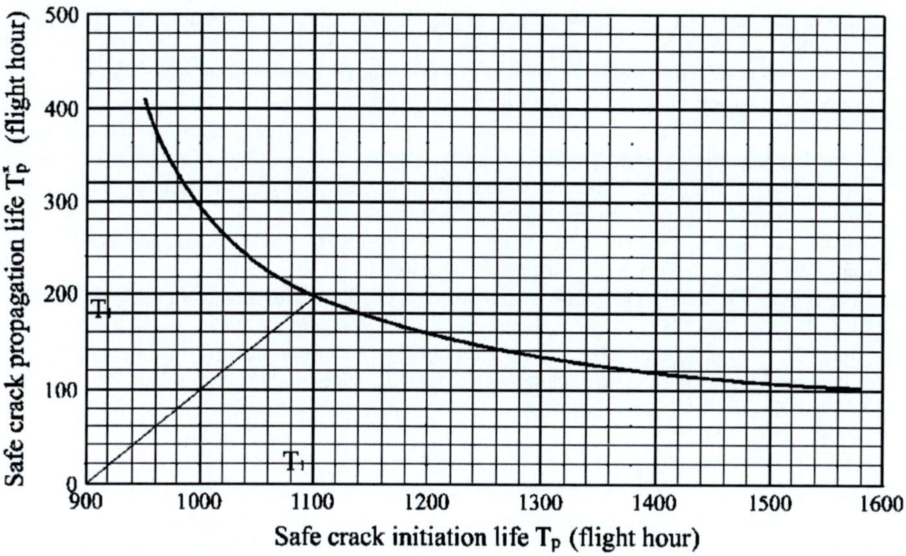

Fig. 5.7 Relationship curve between safe fatigue crack initiation and propagation lives

(or a T–T^* curve of the safe fatigue crack initiation and propagation lives) corresponding to the required reliability level of 0.999999.

Using Eq. 5.93, the relationship curve between the safe fatigue crack initiation and propagation lives can be obtained from Tables 5.1 and 5.2 and shown in Fig. 5.7. From Fig. 5.7, it is easy to obtain the first inspection and maintenance period T_1, namely, when $T_1 = 200$ flight hours, the condition of $T_p = T_p^*$ can be satisfied. As demonstrated in the above application example, using the durability model incorporating safe life methodology and damage tolerance approach, the first inspection and maintenance period can be obtained realistically and easily.

References

1. Committee on Fatigue and Fracture Reliability of the Committee on Structural Safety and Reliability of the Structural Division (1982) Fatigue reliability 1–4. J Struct Division, Proc ASCE 108 ST1:3–88
2. Cheung MMS, Li W (2003) Probabilistic fatigue and fracture analysis of steel bridges. J Struct Saf 23:245–262
3. Liard F (1983) Helicopter fatigue design guide. AGARD-AG-292
4. Grubisic V (1994) Determination of load spectra for design and testing. Int J Veh Des 15:8–26
5. Neugebauer IR, Grubisic V, Fischer G (1989) Procedure for design optimization and durability life approval of truck axles and axle assemblies. SAE Paper 892535, Society of Automotive Engineers, Warrendale, PA
6. Gao ZT, Xiong JJ (2000) Fatigue reliability. Beijing University of Aeronautics and Astronautics Press, Beijing, pp 366–372

7. Freudenthal AM, Garrelts M, Shinozuka M (1966) The analysis of structural safety. J Struct Div ASCE 92:267–325
8. Stewart MG (2001) Reliability-based assessment of ageing bridges using risk ranking and life cycle cost decision analyses. Reliab Eng Syst Saf 74:263–273
9. Tang J, Zhao J (1995) A practical approach for predicting fatigue reliability under random cyclic loading. Reliab Eng Syst Saf 50:7–15
10. Ling J, Pan J (1997) An engineering method for reliability analyses of mechanical structures for long fatigue lives. Reliab Eng Syst Saf 56:135–142
11. Mahadevan S, Dey A, Tryon R, Wang Y, Rousseaus C (2001) Reliability analysis of rotorcraft composite structures. J Aerosp Eng 14(4):140–146
12. Ling J, Gao ZT (1989) Two-dimensional stress-strength interference model. Mech Strength 11(4):12–15
13. Xiong JJ, Gao ZT, Gan WM, Rulin Sun (1996) Fracture reliability design method for structure under sready cyclic loading. Acta Mech Solida Sinca 17(4):235–238
14. Xiong J, Shenoi RA, Gao Z (2002) Small sample theory for reliability design. J Strain Anal Eng Des 37(1):87–92
15. Eurocode 3 (2001) Design of steel structures. British Standards Institution
16. Luo CY, Xiong JJ, Shenoi RA (2008) A reliability-based approach to assess fatigue behaviour based on small incomplete data sets. Adv Mater Res 44–46:871–878
17. Guedes Soares C, Ivanov LD (1989) Time-dependent reliability of the primary ship structure. Reliab Eng Sys Saf 26:59–71
18. Rouhan A, Schoefs F (2003) Probabilistic modeling of inspection results for offshore structures. Struct Saf 25(4):379–399
19. Guedes Soares C, Garbatov Y (1996) Fatigue reliability of the ship hull girder accounting for inspection and repair. Reliab Eng Sys Saf 51:341–351
20. Garbatov Y, Guedes Soares C (2001) Cost and reliability based strategies for fatigue maintenance planning of floating structures. Reliab Eng Sys Saf 73(3):293–301
21. Xiong JJ, Shenoi RA (2009) A durability model incorporating safe life methodology and damage tolerance approach to assess first inspection and maintenance period for structures. Reliab Eng Syst Saf 94:1251–1258
22. Packman PF, Pearson HS, Owens JS, Young G (1969) Definition of fatigue cracks through non-destructive testing. J Mater 4(3):666–700
23. Gao ZT, Jiang XT, Xiong JJ, Guo GH, Gan WM, Xia QY, Wang SP, Zeng BY (1999) Test design and data treatment on fatigue performance–fatigue and fracture performance handbook of metallic materials on helicopter. Beihang University Press, Beijing

Chapter 6
Reliability Prediction for Fatigue Damage and Residual Life in Composites

6.1 Introduction

Static and fatigue behaviour of high performance composites has been investigated extensively. Comprehensive reviews of the subject have been conducted by Reifsnider [1], Talreja [2] and Read and Shenoi [3]. The static tensile strength of composites is governed largely by the fibre strength but this cannot readily be expressed in a simple way because of the statistical nature of the fibre fracture and owing to the role of the matrix and/or interface during fracture. To cause overall composite failure it is generally accepted that there needs to be a critical cluster of fibre breaks adjacent to one another to trigger a catastrophic fracture [4]. The formation of such a cluster is dependent on the statistical distribution of the fibre strengths and the local stress transfer in the vicinity of a fibre break, which in turn is influenced by the fibre-matrix interface. The compression strength of the composite, on the other hand, is related to the ability of the matrix to support the fibre against buckling and the integrity of the fibre-matrix interface. Soutis et al. [5] report compression strength data for high-strength and intermediate modulus fibres in epoxy resins of low to moderate toughness.

During cyclic loading of notched laminates (below the static strength), the initial damage which develops at the notch is similar to that seen under quasistatic loading. This effectively leads to an increase in residual strength with cycling, in particular under tension-tension loading. Under compression loading, the situation may be similar, with the residual strength generally increasing with cycling as a result of the damage growth and an associated reduction in stress concentration associated with the discontinuity. Situations in which fatigue failure of the notched laminates can occur are for the lay-ups where the progressive delamination growth in fatigue leads to a progressively greater instability of the load-carrying 0° plies or ply blocks. This is most likely to be an issue under tension–compression loading. Wang et al. [6] studied notch fatigue strengthening under different cyclic stress levels and elapsed number of cycles in $[0/90]_{4S}$ AS4/PEEK laminates. Nuismer

J. J. Xiong and R. A. Shenoi, *Fatigue and Fracture Reliability Engineering*,
Springer Series in Reliability Engineering, DOI: 10.1007/978-0-85729-218-6_6,

et al. [7] studied uniaxial failure of composite laminates containing stress concentrations. The relationships between the residual strength, fatigue cycles and stress concentration were given. Xiong et al. [8] have undertaken experimental investigations of the static and fatigue strength of T300/QY8911 carbon fibre reinforced composite laminates. It is observed that all tension-tension fatigue damage patterns of notched laminates are similar while the tension-tension fatigue properties vary with the laminate lay-up. The reduction of the stress concentration caused by the tension-tension fatigue damage leads to an improvement in the residual strength of the notched specimens in contrast to their static strength properties. The damage mechanics of notched laminates under compression-compression loading are more complex than those of tension-tension fatigue specimens; their damage patterns are influenced by the test clamping fixture, the layup, the size of specimens and the diameter of hole. The residual strengths are lower than those of the specimens without the fatigue damage. Fatigue strength of composites has been proved to be related to stress ratio (R). Harris et al. [9] have studied the fatigue behaviour of a number of carbon-fibre-reinforced plastics (CFRP) laminates of HTA/913, HTA/982A, T800/5245 and T800/924. Replicate stress/life data were obtained at four stress ratios (R) of +0.1, -0.3, -1.5 and +10 on virgin samples and on damaged samples. The fatigue behaviour of a series of unidirectional hybrid composites has been established as a function of composition and of the stress ratio in repeated tension and tension/compression cycling.

Under compression, as in the static case, fatigue behaviour is more dependent on the matrix and interfacial characteristics than the fibre strength. For [0/90/0] cross ply and [0/90/$\pm$45] quasi-isotropic laminates, damage is of four main types: matrix cracking, delamination in [0/90/$\pm$45] laminates, fibre/matrix debonding and 0° fibre breakage [2]. Because of this variety of fatigue mechanisms closely related to anisotropy, including various mechanisms such as the interfacial debonding, matrix cracking, ply cracking, fiber breakage, and so on, it is difficult to define composite damage in a unique manner. Consequently, a variety of fatigue damage modelling techniques have been proposed that involve stiffness degradation, residual strength, crack density, crack length, etc., with the fatigue damage definition functions being dependent on the number of cycles and on material characteristic variables [10]. While the origins of the damage modelling trends were linear, many researches have proposed non-linear quadratic functions to account for the inherent complexities in composite behaviour. Based on the damage variables mentioned above, various theoretical, experimental and computational criteria have been proposed to estimate the overall life of a composite structure. Most of the damage criteria are functions of the material characteristic variables; damage criteria based on the physical characteristic of the materials have proved to be more promising [11–14]. It is specially emphasized that the damage criteria based on residual strength, stiffness and strain have been extensively developed and widely applied. Because of the simplicity and wide technological usage of the well-known Palmgren–Miner or Paris-Erdogan laws, very precise, experimentally-based, deterministic, linear residual strength or stiffness degradation models have been proposed to predict fatigue life of different laminates from the assumption that the matrix is the weak link of the system.

In recent years, continuum damage mechanics has been applied to investigate the mechanics behavior of the anisotropic composites with multifarious failure modes. In the meso-scale, taking the softening effect of material into account, continuum damage mechanics introduces internal damage variables to denote the formation and growth of micro-cracks and micro-excavations which result in strength and stiffness degradation of material in macro-scale. Jessen and Plumtree [15] derived a damage evolution equation for glass fibre composites pultrusions and the damage S–N curve based on continuum damage principle. Shen et al. [16], Chow and Wang [17], Xiong and Shenoi [18] obtained the elastic damage constitutive relation for composite laminates by means of continuum damage mechanics. Phillips and Shenoi [19, 20] applied damage mechanics to predict fatigue life of structural components such as tee connection and top hat stiffness.

It is interesting in the above reviews to note that a large number of researches are grouped according to some important issues such as static tensile and compressive strength behaviour, tension-tension and compression-compression fatigue behaviour including the influence of the stress ratio on fatigue behaviour, damage mechanisms, and damage models. It is also clear that there is a need for a more practical and expedient model for structural applications, particularly in the aerospace field. In this chapter an attempt is made to develop techniques for modelling the stress- and strain-based residual strengths, to characterise long-term cyclic behaviour and to identify the governing parameters, all to be based on fundamental static and fatigue experimental data.

6.2 Two-Stage Theory on Composite Fatigue Damage

Fatigue process is very complex and involves several damage modes, including fibre/matrix debonding, matrix cracking, delamination, and fibre fracture. In a multi-directional laminate, the 0° fibres carry most of the load and provide most of the stiffness, while the 90° and 45° plies give transverse and shear strength and stiffness, respectively. Under an applied load (mechanical, thermal, static or cyclic) a complicated state of damage develops in the off-axis, causing load re-distributions which lead to fracture in the load-bearing plies. Figure 6.1 shows the various damage mechanisms which take place during in-plane fatigue of a cross-ply laminate, $[0/90]_{ns}$. In fibre composite materials, damage processes are disseminated throughout the sample; there is no equivalent of the single crack that occurs in metals. Failure is gradual and is accompanied by loss of strength and stiffness, or a percentage of the surface covered in visible damage. With the cyclic loading, first there is cracking in the matrix between fibres in the 90° plies in response to the applied loading stresses. These cracks grow until they meet the 0° plies. Individual fibres also fracture randomly with increasing numbers of fatigue cycles. In addition, as the Poisson contraction of the 0° plies is constrained by the

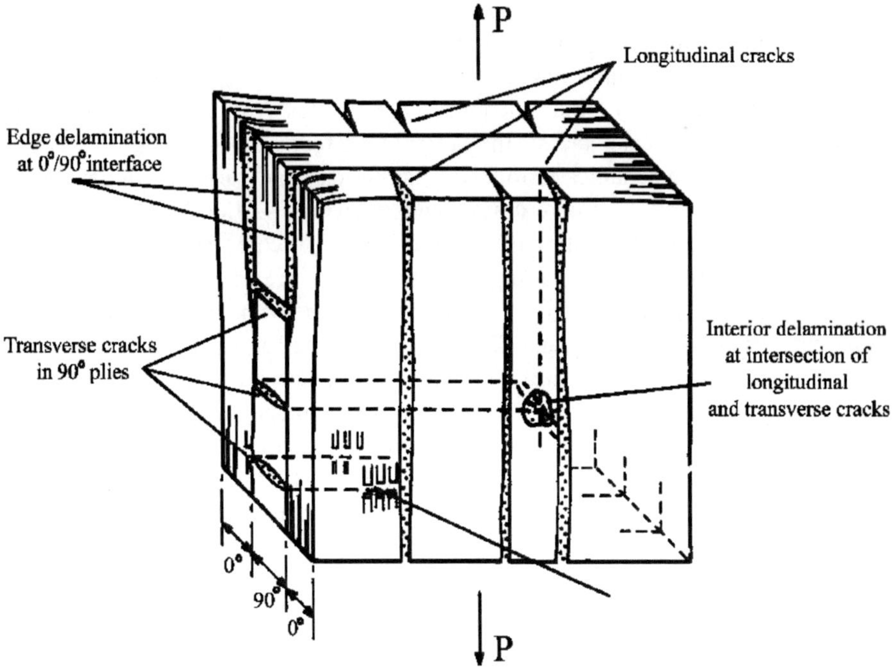

Fig. 6.1 Fatigue damage diagram of composites

90° plies, transverse tensile stresses develop in the 0° plies, causing cracking parallel with the fibres and the applied stress. The longitudinal cracks in the 0° ply are due to the adjacent 90° ply that restrict the Poisson's contraction in the 0° ply. In general layups, matrix cracks parallel to the fibres in the 0° plies occurs at strain close to failure under static loading or may develop throughout fatigue cycling due to the mismatch between Poisson's ratios of adjacent plies. Fibre/matrix debonds initiate around fibres lying at an angle to the loading direction, usually in the 90° ply first, and extend to form microcracks which, in turn, form matrix cracks across the thickness and the width of the ply. The number of cracks in all of the off-axis plies increases with increasing static stress or number of fatigue cycles, reaching a maximum density which remains stable until fracture which depends on ply thickness and orientation and on laminate stacking sequence. Through thickness cracks develop (which is also precipitated by the in-plane matrix cracking) and eventually produce delamination cracks which spread via propagation along the interface between 0° and 90° plies. Delaminations initiate between plies of different orientation and grow inward from the laminate edges across the width of the coupon. The tendency to delaminate results from out-of-plane interlaminar shear and normal stresses which exist at the free-edges and depend on the stacking sequence of the off-axis plies within that laminate. Further interface cracks are produced where the 90° cracks meet the 0° cracks. The cumulative effect of all these processes is a gradual reduction in strength and stiffness.

In short, fatigue damage and failure of composites laminate is revealed at the microscopic level to include the following four stages (as [3]):

(1) The initial stage of damage development consists of multiple matrix cracking along fibres in off-axis plies, culminating in a saturation state of cracking in individual plies. This is called a characteristic damage state (CDS) on the supposition that the state is characteristic of the lay-up and is independent of the load value.
(2) In the stage following CDS the ply cracks link up locally by debonding the ply-ply interface.
(3) Further load cycling causes growth and coalescence of delamination.
(4) The final stage of the damage process is characterized by fibre breakage in the longitudinal plies and total failure.

It is important to note that the exact nature of CDS and the build-up of damage are dependent on material choices and lay-up. Obviously, during Stages (1) and (2), the damage resulting from multiple matrix cracking along fibers in off-axis plies, debonding of the ply-ply interface, etc., is so small that it is very difficult to detect and monitor the damage variable such as the elastic modulus, strain, stiffness reduction, etc. But the damage in Stages (3) and (4) can cause a clear and significant change of the macroscopic properties such as the residual strength, stiffness reduction, etc. of the laminate which fall gradually with increasing numbers of cycles until failure occurs when the residual strength equals the applied cyclic stress.

Therefore, to measure and predict the fatigue life conveniently in engineering design, one can look at damage formation and damage propagation as two basic features. Stages (1) and (2) of fatigue damage of composites represent what is termed as fatigue damage formation, and Stages (3) and (4) together are defined as fatigue damage propagation. In fact, the above analysis and discussion about the relation of damage states and macroscopic properties have been examined to be reasonably consistent with the observations. During fatigue damage formation, the damage has a negligible effect on stiffness change. In fatigue damage propagation, the damage varies, ultimately with a rapidly decaying stiffness. The transition between the two stages is defined as the "damage transition point". Traditional phenomenological fatigue methodology and modern continuum damage mechanics theory are used respectively for fatigue damage formation and propagation life prediction. The two types of prediction method are complementary to each other.

6.3 Fatigue-Driven Residual Strength Model based on Controlling Fatigue Stress

A fatigue damage function is usually proposed as follows [1, 2, 21]:

$$D = D(n, r, S, \omega, T, M, \ldots) \tag{6.1}$$

where D is the damage function, n is the number of cyclic loading, r is the stress ratio, S is the maximum cyclic stress, ω is the loading frequency, T is the current temperature and M is the moisture level. Since the temperature T and moisture level M of the specimen are constants or close to constants (by means of controlling the temperature and moisture rise) during the test. A degenerated form of Eq. 6.1 is also sometimes represented as:

$$D = D(n, r, S, \omega) \tag{6.2}$$

The level of damage can be characterized by residual strength of the specimen or structure, which represents the material strength, usually decreasing with increasing numbers of cycles. In a virgin state, the residual strength equals the static strength; under fatigue loading conditions, the residual strength is equal to the maximum static stress to cause ultimate failure in the post fatigue condition. Thus, the change in residual strength can be regarded as the damage variable. Based on the assumption that the slope of the residual strength $R(n)$ is inversely proportional to some power $b-1$ of the residual strength $R(n)$ itself, a rate equation was obtained as [11]:

$$\frac{dR(n)}{dn} = -\frac{f(r, S, \omega)}{R^{b-1}(n)} \tag{6.3}$$

Here $f(r, S, \omega)$ is a function of r, S and ω. In case of without consideration of loading sequence effect and the change in local stress with damage evolution. Taking the integral of Eq. 6.3 gives:

$$n = f(r, S, \omega)[R_0 - R(n)]^b \tag{6.4}$$

where R_0 is the ultimate strength. Equation 6.4 describes the strength degradation of a sample subjected to constant amplitude, constant frequency cyclic loading. For the sake of simplicity, the loading frequency ω and the stress ratio r will be fixed, so $f(r, S, \omega) = f(S)$. Thus:

$$n = f(S)[R_0 - R(n)]^b \tag{6.5}$$

Equation 6.3 is a surface equation corresponding to residual strength R, fatigue stress S and number of fatigue stress cycle n. On the basis of the S–N curve equation [22] given as:

$$N = C(S - S_0)^m \tag{6.6}$$

where C is the material constant, m is the exponent constant of material, S is the fatigue strength and S_0 is the fitting fatigue limit, or reference fatigue strength, one can obtain $f(S)$ of Eq. 6.6 as:

$$f(S) = C(S - S_0)^m \tag{6.7}$$

Substituting Eq. 6.7 into Eq. 6.3 yields:

$$n = C(S - S_0)^m[R_0 - R(n)]^b \tag{6.8}$$

For a given failure state, namely fatigue residual strength $R(n)$ is given to be a certain of value R_f, Eq. 6.8 becomes S–n curve as $n = C(S - S_0)^m \left[R_0 - R_f\right]^b = C_0(S - S_0)^m$ consistent with S–N curve (6.6) used in fatigue. In the case of given loading stress $S = s_0$, Eq. 6.8 degenerates to the R–n curve as $n = C(s_0 - S_0)^m [R_0 - R(n)]^b = C_0(R_0 - R)^b$, which describes the phenomenological, monotonically quantitatively decreasing law of fatigue residual strength shown in the experiments.

From the randomized approach of deterministic equation presented in Sect. 3.1, a randomization of Eq. 6.8 gives:

$$n = C(S - S_0)^m [R_0 - R(n)]^b X(n) \tag{6.9}$$

where $X(n)$ usually is a log-Gauss stochastic process dependent on n, with 0 mean and constant σ standard deviation. The logarithmic form of Eq. 6.9 is:

$$Y = a_0 + a_1 x_1 + a_2 x_2 + U \tag{6.10}$$

where $Y = \log n$, $a_0 = \log C$, $a_1 = m$, $a_2 = b$, $x_1 = \log(S - S_0)$, $x_2 = \log[R_0 - R(n)]$, $U = \log X(n)$. U is the random variable following a Gauss distribution $N[0, \sigma^2]$. From Eq. 6.10, the random variable Y follows Gauss distribution $N[a_0 + a_1 x_1 + a_2 x_2, \sigma^2]$.

According to the maximum likelihood principle, it can be shown that:

$$a_0 = \bar{y} - a_0 - a_1 \bar{x}_1 - a_2 \bar{x}_2 \tag{6.11}$$

$$a_1 = \frac{L_{12}L_{20} - L_{22}L_{10}}{L_{12}L_{21} - L_{11}L_{22}} \tag{6.12}$$

$$a_2 = \frac{L_{21}L_{10} - L_{11}L_{20}}{L_{12}L_{21} - L_{11}L_{22}} \tag{6.13}$$

$$\sigma = \sqrt{\frac{\sum_{i=1}^{l} (y_i - a_0 - a_1 x_{1i} - a_2 x_{2i})^2}{l}} \tag{6.14}$$

where

$$\bar{y} = \frac{1}{l} \sum_{i=1}^{l} y_i$$

$$\bar{x}_1 = \frac{1}{l} \sum_{i=1}^{l} x_{1i}$$

$$\bar{x}_2 = \frac{1}{l} \sum_{i=1}^{l} x_{2i}$$

$$L_{11} = \sum_{i=1}^{l} (x_{1i} - \bar{x}_1)^2$$

$$L_{22} = \sum_{i=1}^{l} (x_{2i} - \bar{x}_2)^2$$

$$L_{12} = \sum_{i=1}^{l} (x_{1i} - \bar{x}_1)(x_{2i} - \bar{x}_2)$$

$$L_{21} = L_{12}$$

$$L_{10} = \sum_{i=1}^{l} (x_{1i} - \bar{x}_1)(y_i - \bar{y})$$

$$L_{20} = \sum_{i=1}^{l} (x_{2i} - \bar{x}_2)(y_i - \bar{y})$$

Equations 6.11–6.13 are functions of constants R_0 and S_0. The constants R_0 and S_0 can be determined by means of the minimum value principle of the residual sum of squares (RSS) $Q(R_0, S_0)$ in Eq. 6.15 in a four-stage process.

(a) First, letting the residual sum of squares (RSS) as:

$$Q(R_0, S_0) = \sum_{i=1}^{l} (y_i - a_1 - a_2 x_{1i} - a_3 x_{2i})^2 \tag{6.15}$$

(b) Then, the value ranges of R_0 and S_0 are estimated as:

$$R_0 \in (R_{\max}, R_{\max} + \Delta]$$

$$S_0 \in [0, S_{0\,\min})$$

where $R_{\max} = \max\{R_1, R_2, \ldots, R_l\}$, Δ is a finite value, $R_i, (i = 1, 2, \ldots, l)$ is the test data of residual strength, $S_{0\,\min} = \min\{S_1, S_2, \ldots, S_l\}$, and $S_i, (i = 1, 2, \ldots, l)$ is the fatigue stress.

(c) Subsequently, for given initial values of $\hat{R}_0$, $\hat{S}_0$ and calculation step lengths Δ_1, Δ_2, one can search and find out the value of $Q(R_0, S_0)$ by changing the values of R_0 and S_0 during the above value ranges respectively. Thus R_0 and S_0 pertaining to the minimum value of $Q(R_0, S_0)$ may be determined.

(d) Finally, the parameters a_0, a_1 and a_2 can be determined from Eqs. 6.11–6.13.

$$C = 10^{(\bar{y} - a_1 \bar{x}_1 - a_2 \bar{x}_2)}$$

$$m = \frac{L_{12} L_{20} - L_{22} L_{10}}{L_{12} L_{21} - L_{11} L_{22}}$$

$$b = \frac{L_{21}L_{10} - L_{11}L_{20}}{L_{12}L_{21} - L_{11}L_{22}}$$

Using the analogy of Eqs. 3.39 and 3.41 and from Eq. 6.8, the probability models of residual strength based on controlling-stress are derived as:

$$n_p = C(S - S_0)^m [R_0 - R(n)]^b \cdot \exp\left[u_p \hat{k} \sigma\right] \tag{6.16}$$

$$n_{p\gamma} = C(S - S_0)^m [R_0 - R(n)]^b \cdot \exp\left\{\sigma \cdot \left[\hat{k} u_p + t_\gamma \sqrt{\frac{1}{n} + u_p^2 (\hat{k}^2 - 1)}\right]\right\} \tag{6.17}$$

In case there are a large number of test data of ultimate strength R_0, the probability models of residual strength based on controlling-stress can be obtained from the PDF of ultimate strength R_0. According to Eq. 6.5, one can have:

$$R(n) = R_0 - \left[\frac{n}{f(S)}\right]^{\frac{1}{b}} \tag{6.18}$$

Generally, the ultimate strength R_0 follows the two-parameter Weibull distribution [11] and its distribution function is:

$$F_{R_0}(x) = P[R_0 \le x] = 1 - \exp[-(x/\beta)^\alpha] \tag{6.19}$$

Substituting for R_0, the probability distribution function of residual fatigue strength $R(n)$ at a given fatigue stress S and fatigue stress cycles n is

$$F_{R(n)}(x) = P[R(n) \le x] = P\left[R_0 - \left[\frac{n}{f(S)}\right]^{\frac{1}{b}} \le x\right] \tag{6.20}$$

$$F_{R(n)}(x) = P\left\{R_0 \le x + \left[\frac{n}{f(S)}\right]^{\frac{1}{b}}\right\} = 1 - \exp\left\{-\left[\frac{x + \left[\frac{n}{f(S)}\right]^{\frac{1}{b}}}{\beta}\right]^\alpha\right\} \tag{6.21}$$

From Eq. 6.21, it is clear that residual strength $R(n)$ follows a three-parameter Weibull distribution. Based on Eq. 6.18, the residual strength $R_p(n)$ corresponding to a reliability level p is:

$$R_p(n) = R_{0p} - \left[\frac{n}{f(S)}\right]^{\frac{1}{b}} \tag{6.22}$$

Substituting Eq. 6.7 into Eqs. 6.21 and 6.22 gives:

$$F_{R(n)}(x) = P[R(n) \le x] = 1 - \exp\left\{-\left[\frac{x + \left[\frac{n}{C(S-S_0)^m}\right]^{\frac{1}{b}}}{\beta}\right]^\alpha\right\} \tag{6.23}$$

$$R_p(n) = R_{0p} - \left[\frac{n}{C(S - S_0)^m}\right]^{\frac{1}{b}} \tag{6.24}$$

Assuming fatigue failure occurs when the residual strength $R(n)$ equals the maximum cyclic stress S, or $R(n) = S$, at the same time, $n = N$, then a transformation of Eq. 6.8 gives:

$$N = f(S)(R_0 - S)^b \tag{6.25}$$

Since the ultimate strength R_0 is a variable and follows Eq. 6.19, fatigue life N also is a variable and follows:

$$F_N(n) = P[N \le n] = P\left[f(S)(R_0 - S)^b \le n\right] \tag{6.26}$$

$$F_N(n) = P\left\{R_0 \le \left[\frac{n}{f(S)}\right]^{\frac{1}{b}} + S\right\} = 1 - \exp\left\{-\left[\frac{n^{\frac{1}{b}} + S \cdot f^{\frac{1}{b}}(S)}{\beta \cdot f^{\frac{1}{b}}(S)}\right]\right\}^{\alpha} \tag{6.27}$$

Equation 6.27 shows that fatigue life N follows a three-parameter Weibull distribution. When $f(S) = C(S-S_0)^m$, Eq. 6.27 becomes:

$$F_N(n) = P[N \le n] = 1 - \exp\left\{-\left[\frac{n^{\frac{1}{b}} + C^{\frac{1}{b}} \cdot S \cdot (S - S_0)^{\frac{m}{b}}}{\beta \cdot C^{\frac{1}{b}}(S - S_0)^{\frac{m}{b}}}\right]\right\}^{\alpha} \tag{6.28}$$

Example 6.1 37 specimens was cut from $(0, 90, \pm 45)_s$ Graphite/Epoxy laminate panel. 12 specimens were tested statically and their ultimate strengths are shown in Table 6.1. 25 specimens were subjected to fatigue loading at various maximum stress levels s and the results are tabulated in Table 6.2. One specimen did not fail at one million cycles and its residual strength was measured. The fatigue tests were then stopped and the residual strengths were measured. All the fatigue tests were performed at a frequency of 20 Hz with a stress ratio 0.1.

From the test data shown in Tables 6.1 and 6.2, using the model in Eq. 6.8 and its parameter estimation method presented above, the fatigue residual strength surface equation is determined as:

$$n = 4.21 \times 10^{65} S^{-24.56} [519.89 - R(n)]^{0.97} \tag{6.29}$$

From the data shown in Table 6.1, it can be shown that the percentile value pertinent to the probability of survival of 99.9% is 299.31 MPa. Based on

Table 6.1 Ultimate strength of G/E $[0, 90, \pm 45]_s$ (unit: MPa) [11]

435.49	500.82	552.03
457.28	517.53	560.80
495.81	536.12	563.68
498.73	540.05	580.31

Table 6.2 Fatigue residual strength [11] data of G/E [0, 90, ± 45]$_s$

S/MPa	n/cycles	$R(n)$/MPa	S/MPa	n/cycles	$R(n)$/MPa
441.90	1650	441.90	363.92	18790	363.92
441.90	1950	441.90	363.92	3840	363.92
441.90	1320	441.90	348.32	161000	348.32
415.91	2050	415.91	348.32	110000	348.32
389.92	50980	389.92	337.93	523500	337.93
389.92	6480	389.92	337.93	863200	337.93
363.92	155000	363.92	322.33	1346300	322.33
363.92	228500	363.92	337.93	1007000	337.93
363.92	88000	363.92	376.92	30000	376.92
363.92	117580	363.92	376.92	30000	376.92
363.92	228700	363.92	376.92	30000	376.92
363.92	221200	363.92	376.92	30000	376.92
363.92	310000	363.92			

Eq. 6.24, fatigue residual strength surface equation with the probability of survival of 99.9% is:

$$n = 4.21 \times 10^{65} S^{-24.56} [299.31 - R(n)]^{0.97} \tag{6.30}$$

If $S = 363.92$ MPa, Eqs. 6.29 and 6.30 became respectively (shown in Fig. 6.1) as:

$$n = 522.88 \times [519.89 - R(n)]^{0.97} \tag{6.31}$$

$$n = 522.88 \times [299.31 - R(n)]^{0.97} \tag{6.32}$$

Using Eq. 6.31, the mean value of the fatigue loading cycle at a residual strength, $R = 363.92$ MPa is 70102 cycles. From the experimental data shown in Table 6.2, the mean value of fatigue loading cycle at the residual strength of $R = 363.92$ MPa is 152401 cycles. The relative deviation of prediction from the experimental result is:

$$\frac{|152401 - 70102|}{152401} \times 100\% = 54\%$$

If $S = 337.93$ MPa, Eqs. 6.29 and 6.30 became respectively (shown in Fig. 6.2) as

$$n = 3228.11 \times [519.89 - R(n)]^{0.97} \tag{6.33}$$

$$n = 3228.11 \times [299.31 - R(n)]^{0.97} \tag{6.34}$$

Using Eq. 6.33, the mean value of fatigue loading cycle at the residual strength of $R = 363.92$ MPa is 502594 cycles. From the experimental data shown in Table 6.2, the mean value of fatigue loading cycle at the residual strength of

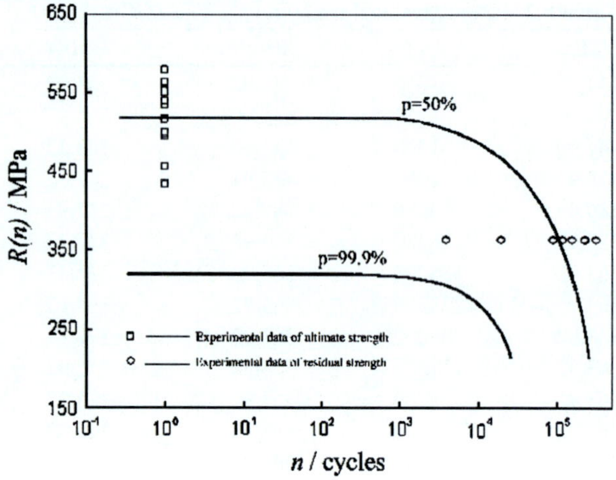

Fig. 6.2 Fatigue residual strength curves (maximum stress $S = 363.92$ MPa)

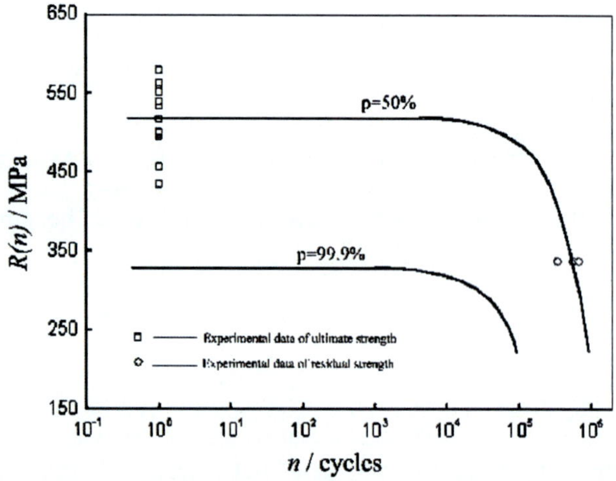

Fig. 6.3 Fatigue residual strength curves (maximum stress $S = 337.93$ MPa)

$R = 363.92$ MPa is 797900 cycles. The relative deviation of prediction from experimental result is

$$\frac{|797900 - 502594|}{797900} \times 100\% = 37\%$$

From Figs. 6.2–6.4, it is seen that the calculated curves are consistent with the experimental data and the probability curves are also realistic. From an accuracy check of the analysis results, the relative deviations of predicted results from experimental data at $S = 363.92$ MPa and $S = 337.93$ MPa are 54 and 37% respectively, with the acceptable scatter. One reason for the deviation of the model results from the experimental data is the small sample size of the test data. It is well known that fatigue lives of composite laminates are often very scattered. For this reason, generally, many sets of large sample experiments are conducted to obtain the population law and the analysis results with high reliability levels.

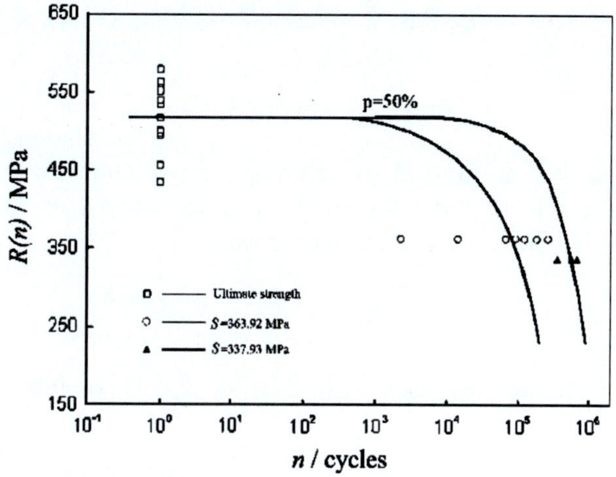

Fig. 6.4 Fatigue residual strength curves

In this case, reasonable constraints limited the experimental results that were feasible and permissible. If more specimens are used for fatigue tests at each stress level and more stress levels are considered, then more exact fatigue performance can be determined and more accurate calculated results can be obtained. An expression in the form of a power function product is adopted in Eq. 6.8 to more easily and expediently estimate the parameters of this model, and there exist four parameters in this model to fit more adequately the experimental phenomenological quantitative laws.

It is worth noting that Eq. 6.8 can characterize fatigue residual strength properties; S and $R(n)$ in this model are expressed in the form of constant amplitude nominal stress, so the loading sequence effect and the change in local stress with damage evolution have not been taken into account.

6.4 Fatigue-Driven Residual Strength Model based on Controlling Fatigue Strain

From Eq. 6.8, the residual strength model based on controlling-stress is:

$$n = C(S - S_0)^m [R_0 - R(n)]^b \tag{6.35}$$

The ultimate strength R_0 and residual strength $R(n)$ are obtained respectively:

$$R_0 = E_0 \cdot \varepsilon_f \tag{6.36}$$

$$R(n) = E(n) \cdot \varepsilon_f \tag{6.37}$$

where ε_f is the rupture strain, E_0 is the initial modulus and $E(n)$ is the residual modulus.

Substituting Eqs. 6.36 and 6.37 into Eq. 6.35, one has:

$$n = C_0(S - S_0)^m[E_0 - E(n)]^b \tag{6.38}$$

where $C_0 = C\varepsilon_f^b$. Equation 6.38 describes the relationship between fatigue stress S, residual modulus $E(n)$ and cyclic loading number n. At a specific fatigue stress S, the relationship between residual modulus $R(n)$ and fatigue strain $\varepsilon(n)$ after n fatigue loading cycles becomes:

$$E(n) = \frac{S}{\varepsilon(n)} \tag{6.39}$$

Substituting Eq. 6.39 into Eq. 6.38, it is possible to have:

$$n = C_0(S - S_0)^m\left[E_0 - \frac{S}{\varepsilon(n)}\right]^b \tag{6.40}$$

Equation 6.40 is the fatigue model based on controlling-strain to depict the relationship between fatigue stress S, fatigue strain $\varepsilon(n)$ and cyclic loading number n. By invoking the procedures of Eqs. 6.11–6.15, one can obtain the undetermined constants E_0, S_0, C_0, m, b and σ in Eq. 6.40. Again, using the analogy of Eqs. 3.39 and 3.41 and from Eq. 6.40, the probability models of residual strength based on controlling-strain are derived as:

$$n_p = C_0(S - S_0)^m\left[E_0 - \frac{S}{\varepsilon(n)}\right]^b \cdot \exp\left[u_p\hat{k}\sigma\right] \tag{6.41}$$

$$n_{p\gamma} = C_0(S - S_0)^m\left[E_0 - \frac{S}{\varepsilon(n)}\right]^b \cdot \exp\left\{\sigma \cdot \left[\hat{k}u_p + t_\gamma\sqrt{\frac{1}{n} + u_p^2\left(\hat{k}^2 - 1\right)}\right]\right\} \tag{6.42}$$

Example 6.2 A 20-ply T300/QY8911 plate was manufactured with a layup of $[45/0_2/-45/90_2/-45/0/45/90]_s$ and having a mean modulus E_0 of 54.34 GPa. Test-specimens were cut from this plate for a scenes of static and fatigue tests. The geometry and dimensions of the specimens are shown in Fig. 6.5. The four variables in the tension case are the dimensions of the central holes φ (10.00 mm),

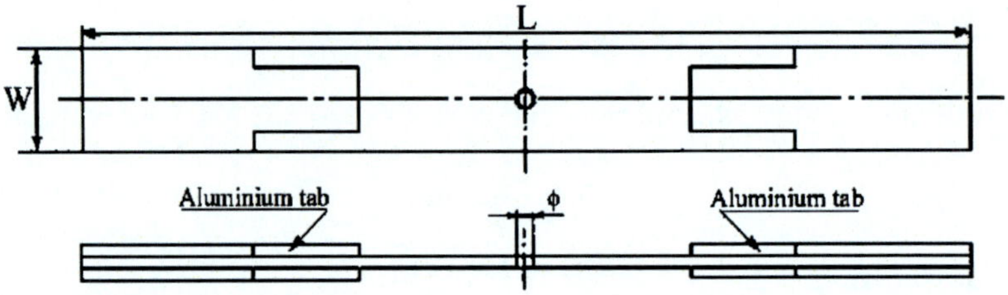

Fig. 6.5 Notched specimen for static tension and tension-tension fatigue tests (all dimensions in mm)

Table 6.3 Strains at the hole-site of notched laminates under tension-tension fatigue loading ($\mu\varepsilon$)

Nominal stress S/MPa	Stress cycles n/cycles				
	1	0.25×10^6	0.5×10^6	0.75×10^6	1.0×10^6
285	4350	4380	4420	4470	4550
	4460				4670
	4490	4500	4520	4570	4760
304	4650				4970
	4620	4680	4730	4800	4920
	4510	4560	4620	4670	4730
320	4930	5180	5270	5310	5350
	4860		5210		5290
	4970				5400
340	5400	$n = 1 \times 10^6$, $\varepsilon = 5780$; $n = 2.2 \times 10^6$, $\varepsilon = 5960$			
	5320	$n = 1 \times 10^6$, $\varepsilon = 5630$; $n = 3.2 \times 10^6$, $\varepsilon = 6030$			

the thickness t (2.4 mm), the width W (50 mm) and the length L (300 mm). 10 specimens were tested statically and the mean value of their fracture strains is 5286 $\mu\varepsilon$. 11 specimens were subjected to tension-tension fatigue loading at various maximum stress levels s and the results are tabulated in Table 6.3. The fatigue tests were performed in a frequency range of 10–15 Hz with a stress ratio of 0.1.

From the test data shown in Table 6.3, by invoking Eqs. 6.11–6.15 and the procedures served from those, one can obtain the undetermined constants E_0, S_0, C_0, m and b in Eq. 6.40. The controlling-strain fatigue residual strength model is then determined as:

$$n = 8.75 \times 10^{30} S^{-19.55} \left[54340.0 - \frac{S}{\varepsilon} \right]^{6.20} \tag{6.43}$$

From Eq. 6.43, it is found that if $S = 285$ MPa, then:

$$n = 9.12 \times 10^{-18} \times \left[54340.0 - \frac{285.0}{\varepsilon} \right]^{6.20} \tag{6.44}$$

If $S = 304$ MPa, then:

$$n = 2.58 \times 10^{-18} \times \left[54340.0 - \frac{304.0}{\varepsilon} \right]^{6.20} \tag{6.45}$$

If $S = 320$ MPa, then:

$$n = 9.48 \times 10^{-19} \times \left[54340.0 - \frac{320.0}{\varepsilon} \right]^{6.20} \tag{6.46}$$

If $S = 340$ MPa, then:

$$n = 2.90 \times 10^{-19} \times \left[54340.0 - \frac{340.0}{\varepsilon}\right]^{6.20} \tag{6.47}$$

Equations 6.43–6.47 are shown in Figs. 6.6–6.10. One can see that there is good agreement between the experimental data and the predicted curves; thus it is argued that the residual fatigue strength model of Eq. 6.40 has adequately and logically characterized the physical characteristics and the phenomenological quantitative laws.

Example 6.3 Two types of 16-ply and 18-ply T300/QY8911 notched specimens were respectively manufactured with the layups of $[45/0/\text{-}45/90]_{2S}$ and $[45/0/\text{-}45/0/90/0/\text{-}45/0/45]_S$ and represented by the notations A and B. The geometry and dimensions of these specimens are shown in Fig. 6.5. The three variables in the tension case are the dimensions of the central holes φ (6.35 mm, 6.35 mm, 8.0 mm and 10.00 mm), the width W (32 mm, 38 mm, 40 mm and 50 mm) and the length L (300 mm, 300 mm, 300 mm and 300 mm). The shoulders of all specimens were bonded with aluminium tabs to suppress the influence of the exposed fibre ends (shown in Fig. 6.5). The aluminium tabs were fabricated into 'staircase' planes to minimize stress concentrations in the specimens at the roots of the shoulders. Three specimens A and B each were tested to determine the ultimate strengths of 336.3 and 470.3 MPa respectively. Fatigue residual strength tests for 11 specimens A and 18 specimens B were conducted under tension–tension (T–T) cyclic loadings of constant stress amplitude at room temperature and moisture by using a stress ratio of $R = 0.1$ and a loading frequency of 15 Hz. The experiments are listed in Table 6.4 shows that the fatigue residual strengths of Laminate A after 1×10^6 cycles are also close to each other, and this is also true for specimen B.

Fig. 6.6 Controlling-strain residual strength surface for laminate $[45/0_2/-45/90_2/-45/0/45/90]_s$ of T300/QY8911

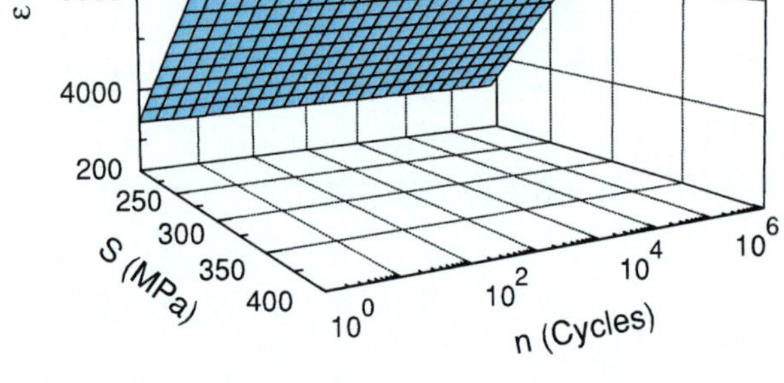

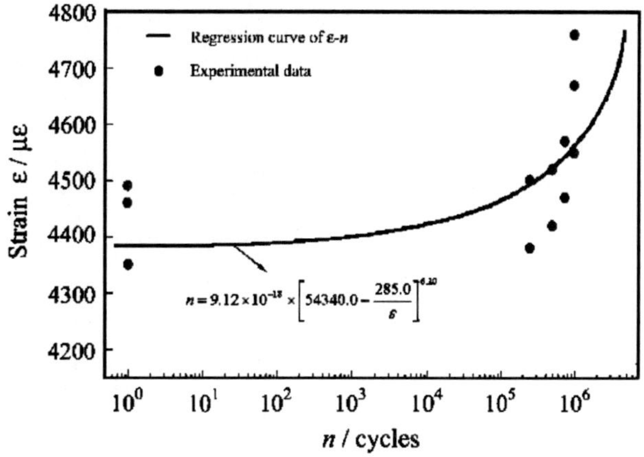

Fig. 6.7 Fatigue experimental data and the fitting strain-life curve for laminate $[45/0_2/-45/90_2/-45/0/45/90]_s$ of T300/QY8911 at $S = 285$ MPa

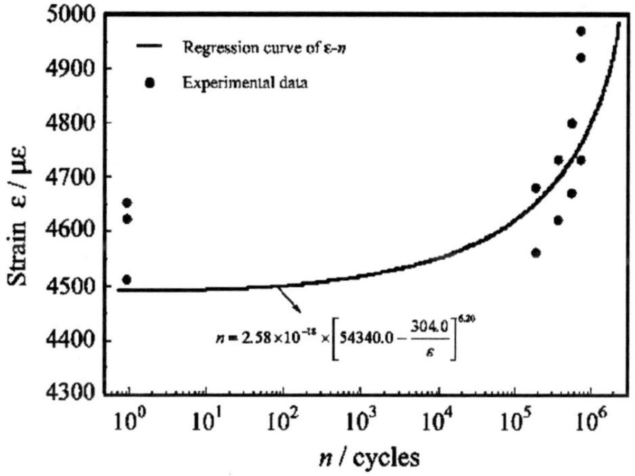

Fig. 6.8 Fatigue experimental data and the fitting strain-life curve for laminate $[45/0_2/-45/90_2/-45/0/45/90]_s$ of T300/QY8911 at $S = 304$ MPa

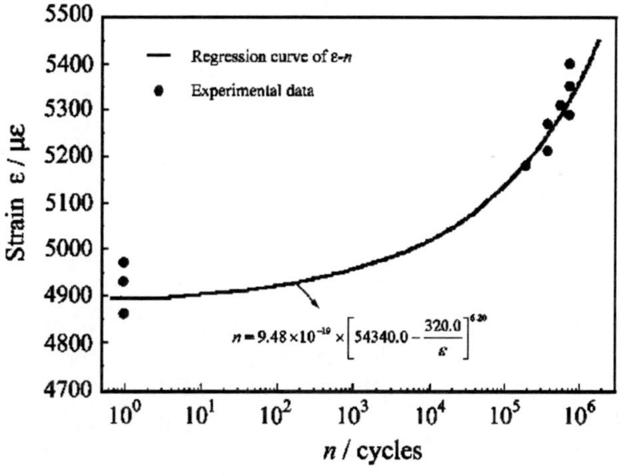

Fig. 6.9 Fatigue experimental data and the fitting strain-life curve for laminate $[45/0_2/-45/90_2/-45/0/45/90]_s$ of T300/QY8911 at $S = 320$ MPa

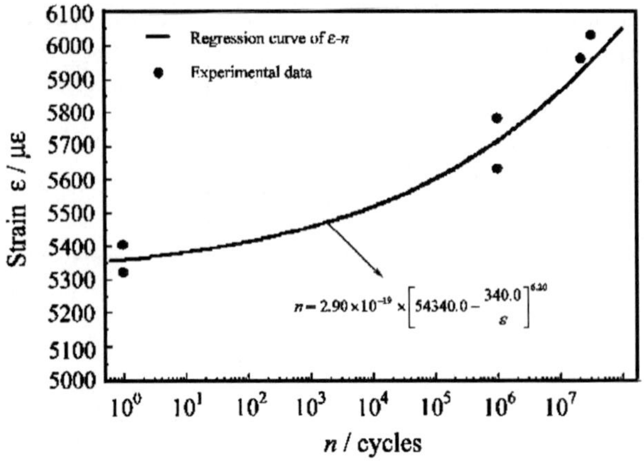

Fig. 6.10 Fatigue experimental data and the fitting strain-life curve for laminate $[45/0_2/-45/90_2/-45/0/45/90]_s$ of T300/QY8911 at $S = 340$ MPa

Table 6.4 Strains at the hole-site of notched Laminates A and B under tension-tension fatigue loading ($\mu\varepsilon$)

No.	Nominal stress *S*/MPa	Stress cycles *n*/cycles					Residual strength *R*(*n*)/MPa
		1	0.25×10^6	0.5×10^6	0.75×10^6	1.0×10^6	
A-1	294	4625	4625	4625	4625	4625	395.8
A-2	301	4580	4590	4590	4620	4760	411.8
A-3	291	4530	4530	4540	4560	4570	381.9
A-4	318	4850	4900	4975	5190	5310	403.8
A-5	312	4610	5050	5090	5110	5170	387.4
A-6	324	4875	4970	5020	5080	5160	409.5
A-7	318	4900	5010	5040	5180	5470	–
A-8	319	4850	4990	5020	5030	5280	–
A-9	334	5180	5330	5410	5620	6010	408.6
A-10	330	5210	5620	5690	5780	6120	401.3
A-11	341	5270	5530	5580	5610	6080	377.4
B-1	370	4000	4000	4000	4000	4000	511.0
B-2	383	4040	4040	4040	4050	4080	530.2
B-3	386	4020	4040	4040	4040	4060	529.4
B-4	412	4400	4400	4400	4420	4450	531.6
B-4	408	4750	4750	4750	4750	4750	536.4
B-6	396	4560	4630	4630	4630	4640	505.6
B-7	392	4360	4370	4390	4390	4410	–
B-8	405	4440	4480	4500	4520	4530	–
B-9	425	4670	4720	4750	4790	4920	552.4
B-10	426	4720	4750	4750	4770	4860	523.3
B-11	430	4710	4730	4740	4810	4900	521.8
B-12	441	4790	4800	4810	4830	4920	–
B-13	428	4800	4800	4820	4840	4890	–
B-14	459	5040	5110	5180	5270	5380	541.2
B-15	450	4950	5050	5280	5330	5400	554.5
B-16	458	5010	5090	5240	5260	5310	533.6
B-17	465	5060	5190	5250	5350	5400	–
B-18	469	5050	5110	5170	5310	5420	–

Their mean values after 1×10^6 cycles are respectively calculated as: 397.5 and 530.1 MPa. Comparing the static tensile and the mean values of tensile-tensile fatigue residual strength data of the notched specimens of Laminates A and B after 1×10^6 cycles, one can see that all the tensile-tensile fatigue residual strengths are greater than the static tensile strength. This is because of the reduction of stress concentration caused by tension-tension fatigue damage and which consequently lead to an improvement in the residual strength of the notched specimens, in contrast to their static strength properties.

Similarly, from the experimental results listed in Table 6.4, by analogy of Eqs. 6.11–6.15, one can be obtain the undetermined constants E_0, S_0, C_0, m and b in Eq. 6.40. The controlling-strain fatigue residual strength model for Laminate A is then obtained as:

$$n = 1.48 \times 10^{47} S^{-31.56} \left[55210.0 - \frac{S}{\varepsilon}\right]^{9.54} \tag{6.48}$$

From Eq. 6.48, it is found that if $S = 232.5$ MPa, then:

$$n = 3.24 \times 10^{-19} \times \left[55210.0 - \frac{232.5}{\varepsilon}\right]^{6.77} \tag{6.49}$$

If $S = 249$ MPa, then:

$$n = 4.67 \times 10^{-28} \times \left[55210.0 - \frac{249.1}{\varepsilon}\right]^{9.01} \tag{6.50}$$

If $S = 265.7$ MPa, then:

$$n = 8.65 \times 10^{-38} \times \left[55210.0 - \frac{265.7}{\varepsilon}\right]^{11.24} \tag{6.51}$$

The trends determined from Eqs. 6.48–6.51 are shown in Figs. 6.11–6.14 respectively. Using the analogy of Eqs. 6.48–6.51, one finds the controlling-strain fatigue residual strength model for Laminate B to be:

$$n = 1.79 \times 10^{-6} S^{-3.91} \left[77120.0 - \frac{S}{\varepsilon}\right]^{6.05} \tag{6.52}$$

If $S = 324.9$ MPa, then:

$$n = 1.91 \times \left[77210.0 - \frac{324.9}{\varepsilon}\right]^{1.64} \tag{6.53}$$

If $S = 348.9$MPa, then:

$$n = 2.44 \times 10^{-31} \left[77210.0 - \frac{348.1}{\varepsilon}\right]^{10.09} \tag{6.54}$$

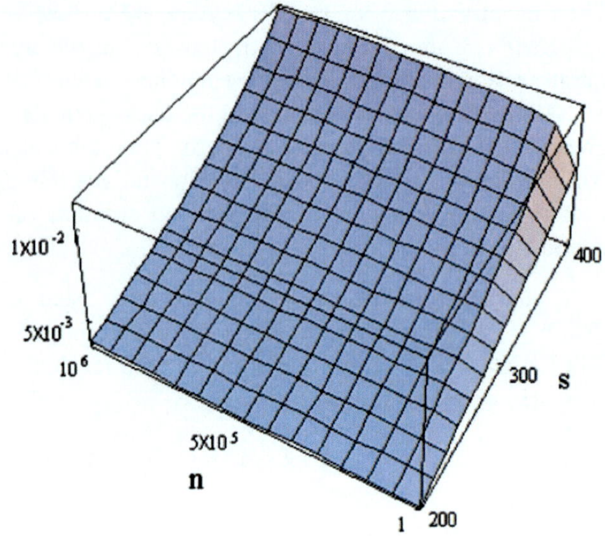

Fig. 6.11 Fatigue residual strength surface for Laminate A

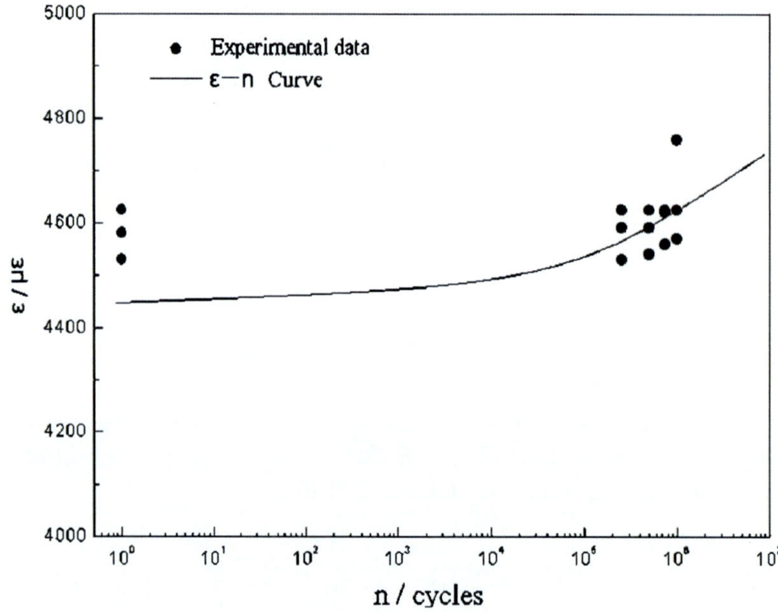

Fig. 6.12 Fatigue experimental data and the fitting strain-life curve for Laminate A at $S = 232.5$ MPa

If $S = 371.3$ MPa, then:

$$n = 6.67 \times^{-35} \left[77210.0 - \frac{371.3}{\varepsilon}\right]^{10.80} \tag{6.55}$$

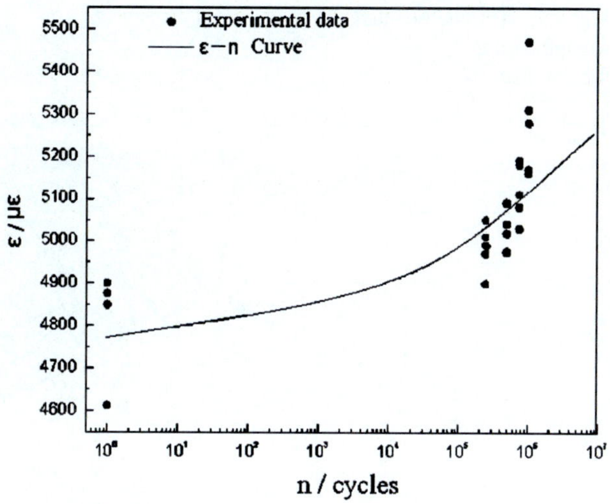

Fig. 6.13 Fatigue experimental data and the fitting strain-life curve for Laminate A at $S = 249$ MPa

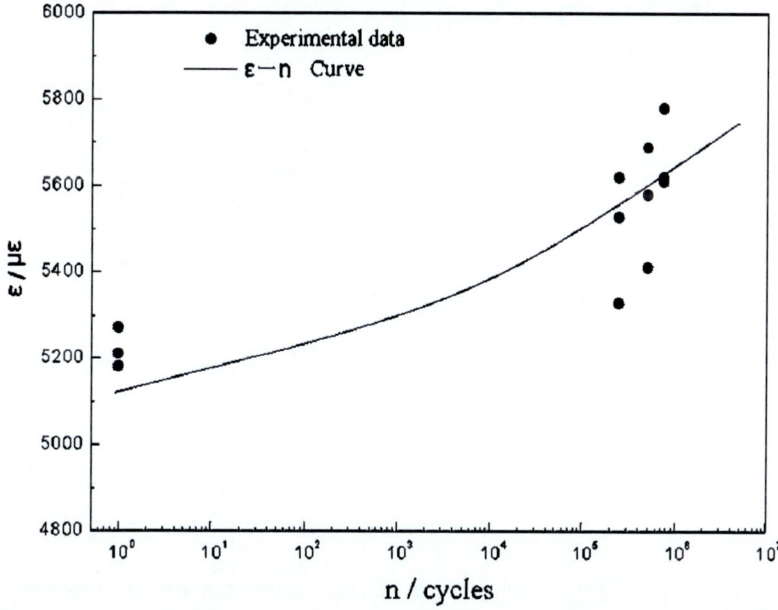

Fig. 6.14 Fatigue experimental data and the fitting strain-life curve for Laminate A at $S = 265.7$ MPa

Equations 6.52–6.55 are shown in Figs. 6.15–6.18. From this, it is evident that there is good agreement between the experimental data and the predicted curves; thus it is argued that the residual fatigue strength model of Eq. 6.40 has adequately and logically characterized the physical characteristics and the phenomenological quantitative laws.

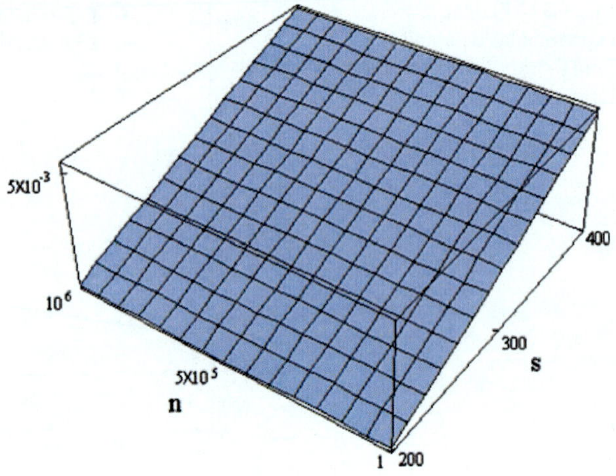

Fig. 6.15 Fatigue residual strength surface for Laminate B

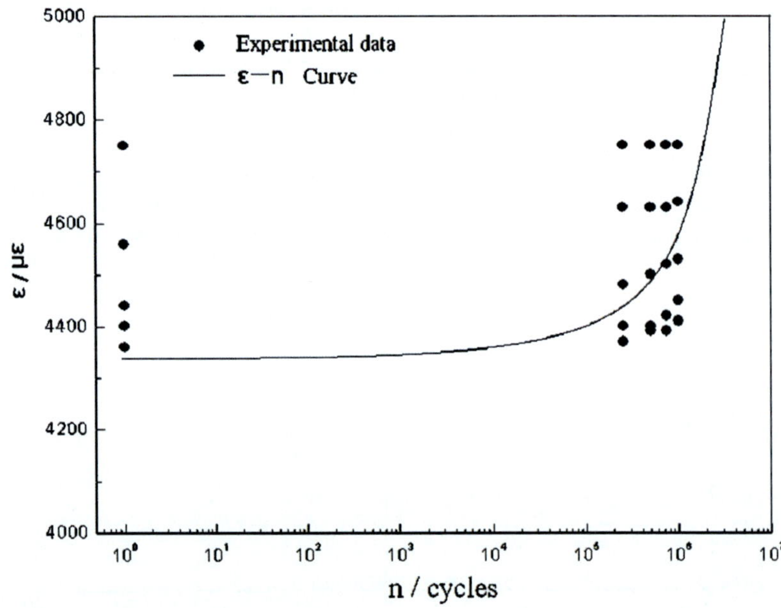

Fig. 6.16 Fatigue experimental data and the fitting strain-life curve for Laminate B at $S = 324.9$ MPa

6.5 Constitutive Relations for Composite Damage

The three-dimensional undamaged constitutive relation for anisotropic composites can be expressed in a general form as $\varepsilon = S{:}\sigma$, where S is the fourth-order compliance tensor with 21 independent parameters and ε is the strain tensor. For orthotropic materials, S has only 9 independent parameters in the case where the major direction of the material coincides with the coordinate axis, and can be written as:

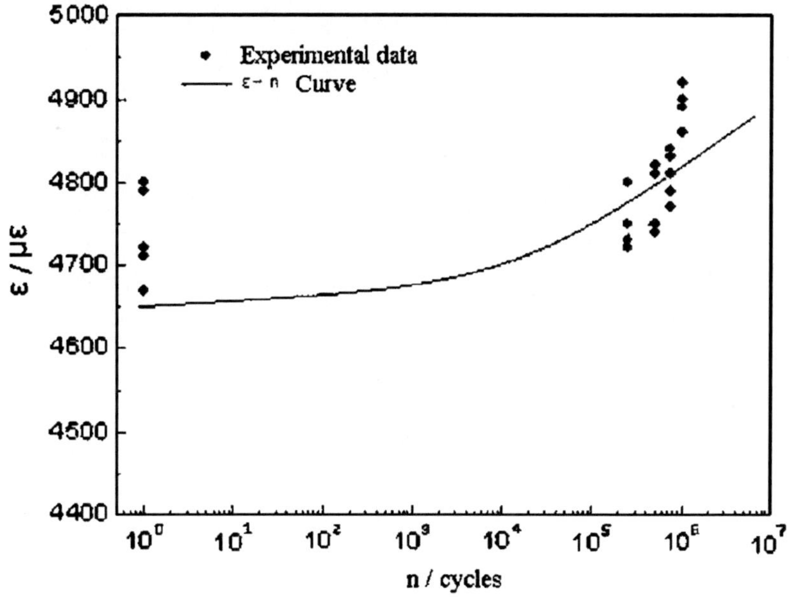

Fig. 6.17 Fatigue experimental data and the fitting strain-life curve for Laminate B at $S = 348.9$ MPa

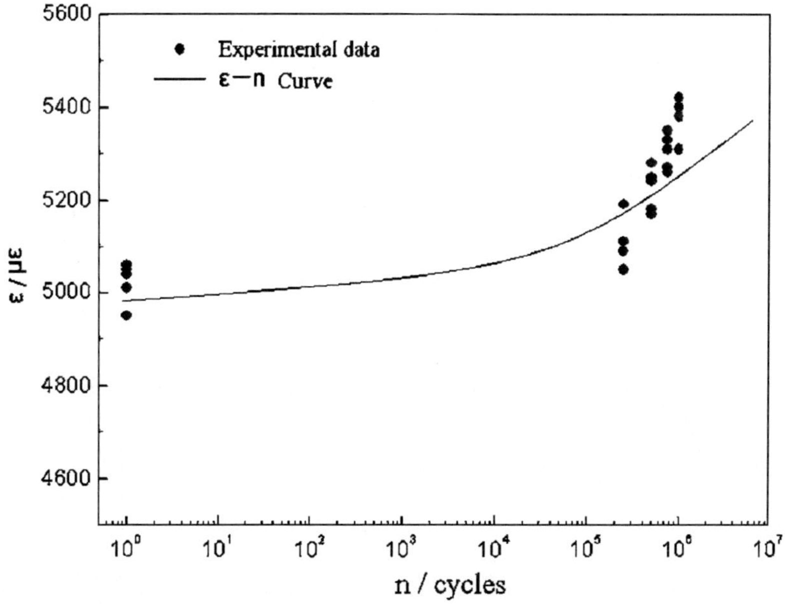

Fig. 6.18 Fatigue experimental data and the fitting strain-life curve for Laminate B at $S = 371.3$ MPa

$$S = \begin{bmatrix} \frac{1}{E_1} & -\frac{\nu_{21}}{E_2} & -\frac{\nu_{31}}{E_3} & 0 & 0 & 0 \\ -\frac{\nu_{12}}{E_1} & \frac{1}{E_2} & -\frac{\nu_{32}}{E_3} & 0 & 0 & 0 \\ -\frac{\nu_{13}}{E_1} & -\frac{\nu_{23}}{E_2} & \frac{1}{E_3} & 0 & 0 & 0 \\ 0 & 0 & 0 & \frac{1}{G_{23}} & 0 & 0 \\ 0 & 0 & 0 & 0 & \frac{1}{G_{31}} & 0 \\ 0 & 0 & 0 & 0 & 0 & \frac{1}{G_{12}} \end{bmatrix} \tag{6.56}$$

where 1, 2, 3 express the three major directions of the material. E is the elastic modulus, G is the shear modulus and ν is the Poisson's ratio. For a transversely isotropic material, $E_1 = E_2$, $\nu_{12} = \nu_{21}$, $\nu_{13} = \nu_{31} = \nu_{23} = \nu_{32}$, $G_{12} = 2\left(\frac{1}{E_1} + \frac{\nu_{12}}{E_1}\right)$; so it is seen from Eq. 6.56 that **S** has only 5 independent parameters in the case where the major direction of the material coincides with the coordinate axis, and can be written as:

$$\mathbf{S} = \begin{bmatrix} \frac{1}{E_1} & -\frac{\nu_{12}}{E_1} & -\frac{\nu_{13}}{E_1} & 0 & 0 & 0 \\ -\frac{\nu_{12}}{E_1} & \frac{1}{E_1} & -\frac{\nu_{13}}{E_1} & 0 & 0 & 0 \\ -\frac{\nu_{13}}{E_1} & -\frac{\nu_{13}}{E_1} & \frac{1}{E_3} & 0 & 0 & 0 \\ 0 & 0 & 0 & \frac{1}{G_{23}} & 0 & 0 \\ 0 & 0 & 0 & 0 & \frac{1}{G_{23}} & 0 \\ 0 & 0 & 0 & 0 & 0 & 2\left(\frac{1}{E_1} + \frac{\nu_{12}}{E_1}\right) \end{bmatrix} \tag{6.57}$$

For an anisotropic elastic composite material in a damaged state, an effective stress tensor $\tilde{\sigma}$ is introduced and has the following relation with the nominal stress tensor σ:

$$\tilde{\sigma} = \boldsymbol{M}(\boldsymbol{D}) : \sigma \tag{6.58}$$

where $\boldsymbol{M}(\boldsymbol{D})$ is a fourth-order damage transformation tensor with 21 independent parameters. When the major axis of the stress coincides with the propagation direction of fatigue damage, $\boldsymbol{M}(\boldsymbol{D})$ can be written as:

$$M_{ijkl}(\boldsymbol{D}) = S_{ij}\delta_{ik}\delta_{jl} \quad (i, j \text{ not summed}) \tag{6.59}$$

A convenient way to define $\boldsymbol{M}(\boldsymbol{D})$, with a view to decrease the internal variables, is as follows:

$$\boldsymbol{M}(\boldsymbol{D}) = \begin{bmatrix} \frac{1}{\sqrt{1-D_1}} & & & & & \\ & \frac{1}{\sqrt{1-D_2}} & & & & \\ & & \frac{1}{\sqrt{1-D_3}} & & & \\ & & & \frac{1}{\sqrt{1-D_4}} & & \\ & & & & \frac{1}{\sqrt{1-D_5}} & \\ & & & & & \frac{1}{\sqrt{1-D_6}} \end{bmatrix} \tag{6.60}$$

where the damage variable D_i is defined as:

$$D_i = \begin{cases} 1 - \frac{\tilde{E}_i}{E_i}, (i = 1, 2, 3) \\ 1 - \frac{\tilde{G}_i}{G_i}, (i = 4, 5, 6) \end{cases} \tag{6.61}$$

and $\tilde{G}_4 = \tilde{G}_{23}$, $\tilde{G}_5 = \tilde{G}_{13}$, $\tilde{G}_6 = \tilde{G}_{12}$. $\tilde{E}$ is the effective elastic modulus, E_0 is the elastic modulus of elasticity in an undamaged state and $\tilde{G}$ is the effective shear modulus.

In terms of the energy equivalence principle [23], the undamaged complementary energy corresponding to the effective stress tensor $\tilde{\sigma}$ is equal to the elastic damaged complementary energy at the nominal stress tensor σ, namely:

$$W_e(\tilde{\sigma}, 0) = W_e(\sigma, \boldsymbol{D}) \tag{6.62}$$

where $\mathbf{D}$ is the damage tensor. $W_e(\tilde{\sigma}, 0)$ is the damaged complementary energy density. $W_e(\sigma, \mathbf{D})$ is the damaged complementary energy density. σ is the nominal stress tensor. $\tilde{\sigma}$ is the effective stress tensor

The undamaged complementary energy density at the effective stress tensor $\tilde{\sigma}$ is:

$$W_e(\tilde{\sigma}, 0) = \frac{1}{2} \tilde{\sigma}^T : S : \tilde{\sigma} \tag{6.63}$$

Substituting Eqs. 6.58 into 6.63 yields:

$$W_e(\tilde{\sigma}, 0) = \sigma^{\mathrm{T}} : \boldsymbol{M}^{\mathrm{T}}(\boldsymbol{D}) : \boldsymbol{S} : \boldsymbol{M}(\boldsymbol{D}) : \sigma \tag{6.64}$$

According to Eqs. 6.62 and 6.64, the damaged complementary energy density is as follows:

$$W_e(\sigma, D) = W_e(\tilde{\sigma}, 0) = \sigma^{\mathrm{T}} : \boldsymbol{M}^{\mathrm{T}}(\boldsymbol{D}) : \boldsymbol{S} : \boldsymbol{M}(\boldsymbol{D}) : \sigma \tag{6.65}$$

Equation 6.65 can be written as:

$$W_e(\sigma, \boldsymbol{D}) = \sigma^{\mathrm{T}} : \boldsymbol{S}(\boldsymbol{D}) : \sigma \tag{6.66}$$

where $\boldsymbol{S}(\boldsymbol{D})$ is the damage compliance and can be determined as:

$$\boldsymbol{S}(\boldsymbol{D}) = \boldsymbol{M}^{\mathrm{T}}(\boldsymbol{D}) : \boldsymbol{S} : \boldsymbol{M}(\boldsymbol{D}) \tag{6.67}$$

Hence, by means of the complementary energy principle, the damage constitutive equation can be obtained as:

$$\varepsilon = \frac{\partial W_e(\sigma, \boldsymbol{D})}{\partial \sigma} = \boldsymbol{S}(\boldsymbol{D}) : \sigma \tag{6.68}$$

Or:

$$\sigma = \boldsymbol{S}^{-1}(\boldsymbol{D}) : \varepsilon \tag{6.69}$$

For orthotropic materials, based on Eqs. 6.56, 6.60 and 6.67, the damage compliance tensor is:

$$S(D)=\begin{bmatrix} \frac{1}{E_1(1-D_1)} & \frac{-\nu_{21}}{E_2\sqrt{(1-D_2)(1-D_1)}} & \frac{-\nu_{31}}{E_3\sqrt{(1-D_3)(1-D_1)}} & 0 & 0 & 0 \\ \frac{-\nu_{12}}{E_1\sqrt{(1-D_1)(1-D_2)}} & \frac{1}{E_2(1-D_2)} & \frac{-\nu_{32}}{E_3\sqrt{(1-D_3)(1-D_2)}} & 0 & 0 & 0 \\ \frac{-\nu_{13}}{E_1\sqrt{(1-D_1)(1-D_3)}} & \frac{-\nu_{23}}{E_2\sqrt{(1-D_2)(1-D_3)}} & \frac{1}{E_3(1-D_3)} & 0 & 0 & 0 \\ 0 & 0 & 0 & \frac{1}{G_{23}(1-D_4)} & 0 & 0 \\ 0 & 0 & 0 & 0 & \frac{1}{G_{31}(1-D_5)} & 0 \\ 0 & 0 & 0 & 0 & 0 & \frac{1}{G_{12}(1-D_6)} \end{bmatrix} \quad (6.70)$$

From Eqs. 6.57, 6.60 and 6.67, the damage compliance tensor for a transversely isotropic material becomes:

$$\begin{bmatrix} \frac{1}{E_1(1-D_1)} & \frac{-\nu_{12}}{E_1\sqrt{(1-D_2)(1-D_1)}} & \frac{-\nu_{13}}{E_1\sqrt{(1-D_3)(1-D_1)}} & 0 & 0 & 0 \\ \frac{-\nu_{12}}{E_1\sqrt{(1-D_1)(1-D_2)}} & \frac{1}{E_1(1-D_2)} & \frac{-\nu_{13}}{E_1\sqrt{(1-D_3)(1-D_2)}} & 0 & 0 & 0 \\ \frac{-\nu_{13}}{E_1\sqrt{(1-D_1)(1-D_3)}} & \frac{-\nu_{13}}{E_1\sqrt{(1-D_2)(1-D_3)}} & \frac{1}{E_3(1-D_3)} & 0 & 0 & 0 \\ 0 & 0 & 0 & \frac{1}{G_{23}(1-D_4)} & 0 & 0 \\ 0 & 0 & 0 & 0 & \frac{1}{G_{23}(1-D_5)} & 0 \\ 0 & 0 & 0 & 0 & 0 & \frac{1}{2(1-D_6)}\left(\frac{1}{E_1}+\frac{\nu_{12}}{E_1}\right) \end{bmatrix} \quad (6.71)$$

According to Eq. 6.70, in a two-dimensional stress state, the damage compliance tensor for the orthotropic material is of the form:

$$S(D)=\begin{bmatrix} \frac{1}{E_1(1-D_1)} & \frac{-\nu_{21}}{E_2\sqrt{(1-D_1)(1-D_2)}} & 0 \\ \frac{-\nu_{12}}{E_1\sqrt{(1-D_1)(1-D_2)}} & \frac{1}{E_2(1-D_2)} & 0 \\ 0 & 0 & \frac{1}{G_{12}(1-D_6)} \end{bmatrix} \quad (6.72)$$

The damage stiffness tensor is:

$$E(D)=S^{-1}(D)=\begin{bmatrix} \frac{E_1(1-D_1)}{1-\nu_{12}\nu_{21}} & \frac{\nu_{21}E_2\sqrt{(1-D_1)(1-D_2)}}{1-\nu_{12}\nu_{21}} & 0 \\ \frac{\nu_{12}E_1\sqrt{(1-D_1)(1-D_2)}}{1-\nu_{12}\nu_{21}} & \frac{E_2(1-D_2)}{1-\nu_{12}\nu_{21}} & 0 \\ 0 & 0 & G_{12}(1-D_6) \end{bmatrix} \quad (6.73)$$

where $E(D)$ is the damage stiffness tensors.

Then, by means of Eqs. 6.66 and 6.72, the damaged complementary energy density for the orthotropic material in a two-dimensional stress state can be written as:

$$W_e(\sigma, \boldsymbol{D}) = \frac{\sigma_1^2}{2E_1(1-D_1)} - \frac{\nu_{12}\sigma_1\sigma_2}{2E_1\sqrt{(1-D_1)(1-D_2)}} - \frac{\nu_{21}\sigma_1\sigma_2}{2E_2\sqrt{(1-D_1)(1-D_2)}} + \frac{\sigma_2^2}{2E_2(1-D_2)} + \frac{\tau_{12}^2}{2G_{12}(1-D_6)} \tag{6.74}$$

Using coordinate transformation, in case the major stress directions are not coincident with the coordinate axis, the compliance and stiffness tensors in the non-elastic major direction can be determined. Thus the damage constitutive relation and evolution equation in the non-elastic major direction are established. Let l_1, m_1, n_1, l_2, m_2, n_2 and l_3, m_3, n_3 be the cosines of the included angles between the coordinate axis x', y', z' and the original coordinate axis x, y, z. The transformation tensors of the stress and strain tensors are respectively:

$$\boldsymbol{A} = \begin{bmatrix} l_1^2 & m_1^2 & n_1^2 & 2m_1n_1 & 2n_1l_1 & 2l_1m_1 \\ l_2^2 & m_2^2 & n_2^2 & 2m_2n_2 & 2n_2l_2 & 2l_2m_2 \\ l_3^2 & m_3^2 & n_3^2 & 2m_3n_3 & 2n_3l_3 & 2l_3m_3 \\ l_2l_3 & m_2m_3 & n_2n_3 & m_2n_3+m_3n_2 & n_2l_3+n_3l_2 & l_2m_3+l_3m_2 \\ l_3l_1 & m_3m_1 & n_3n_1 & m_3n_1+m_1n_3 & n_3l_1+n_1l_3 & l_3m_1+l_1m_3 \\ l_1l_2 & m_1m_2 & n_1n_2 & m_1n_2+m_2n_1 & n_1l_2+n_2l_1 & l_1m_2+l_2m_1 \end{bmatrix} \tag{6.75}$$

$$\boldsymbol{B} = \begin{bmatrix} l_1^2 & m_1^2 & n_1^2 & m_1n_1 & n_1l_1 & l_1m_1 \\ l_2^2 & m_2^2 & n_2^2 & m_2n_2 & n_2l_2 & l_2m_2 \\ l_3^2 & m_3^2 & n_3^2 & m_3n_3 & n_3l_3 & l_3m_3 \\ 2l_2l_3 & 2m_2m_3 & 2n_2n_3 & m_2n_3+m_3n_2 & n_2l_3+n_3l_2 & l_2m_3+l_3m_2 \\ 2l_3l_1 & 2m_3m_1 & 2n_3n_1 & m_3n_1+m_1n_3 & n_3l_1+n_1l_3 & l_3m_1+l_1m_3 \\ 2l_1l_2 & 2m_1m_2 & 2n_1n_2 & m_1n_2+m_2n_1 & n_1l_2+n_2l_1 & l_1m_2+l_2m_1 \end{bmatrix} \tag{6.76}$$

Because $\boldsymbol{A}^{-1} = \boldsymbol{B}^{\mathrm{T}}$, $\boldsymbol{B}^{-1} = \boldsymbol{A}^{\mathrm{T}}$, the compliance and stiffness tensors in the non-elastic major direction are respectively:

$$\boldsymbol{S}' = \boldsymbol{B} : \boldsymbol{S} : \boldsymbol{B}^{\mathrm{T}} \tag{6.77}$$

$$\boldsymbol{E}' = \boldsymbol{A} : \boldsymbol{E} : \boldsymbol{A}^{\mathrm{T}} \tag{6.78}$$

where $\boldsymbol{S}'(\boldsymbol{D})$ is the damage compliance tensors in non-elastic major direction and $\boldsymbol{E}'(\boldsymbol{D})$ is the damage stiffness tensors in non-elastic major direction.

Then the damage compliance and stiffness tensors in the non-elastic major direction take the form as:

$$\boldsymbol{S}'(\boldsymbol{D}) = \boldsymbol{B} : \boldsymbol{S}(\boldsymbol{D}) : \boldsymbol{B}^{\mathrm{T}} \tag{6.79}$$

$$\boldsymbol{E}'(\boldsymbol{D}) = \boldsymbol{A} : \boldsymbol{E}(\boldsymbol{D}) : \boldsymbol{A}^{\mathrm{T}} \tag{6.80}$$

6.6 Stress Concentration of Notched Anisotropic Laminate

If an infinite anisotropic laminated plate containing an elliptic hole with the half axis lengths of a and b is subjected to a nominal stress σ_0 whose direction has an angle of α with the long axis OX of the elliptic hole (shown in Fig. 6.19), the stress along the elliptic hole of the anisotropic laminate is given as [24]:

$$\sigma_x = -\sigma_0 \cos^2 \alpha + 2\mathrm{Re}\left[s_1^2 \phi_0(z_1) + s_2^2 \psi_0(z_2)\right] \tag{6.81}$$

$$\sigma_y = -\sigma_0 \sin^2 \alpha + 2\mathrm{Re}[\phi_0(z_1) + \psi_0(z_2)] \tag{6.82}$$

$$\tau_{xy} = -\sigma_0 \sin \alpha \cos \alpha - 2\mathrm{Re}[s_1 \phi_0(z_1) + s_2 \psi_0(z_2)] \tag{6.83}$$

where σ_x is the normal stress at x direction in plane state of stress. σ_y is the normal stress at y direction in plane state of stress and τ_{xy} is the shear stress in plane state of stress:

$$\phi_0(z_1) = \frac{i\sigma_0(a - is_1 b)}{4(s_1 - s_2)} \left\{ \frac{b(s_2 \sin 2\alpha + 2\cos^2 \alpha)}{z_1 + \sqrt{z_1^2 - (a^2 + s_1^2 b^2)}} + \frac{ia(2s_2 \sin^2 \alpha + \sin 2\alpha)}{z_1 + \sqrt{z_1^2 - (a^2 + s_1^2 b^2)}} \right\} \tag{6.84}$$

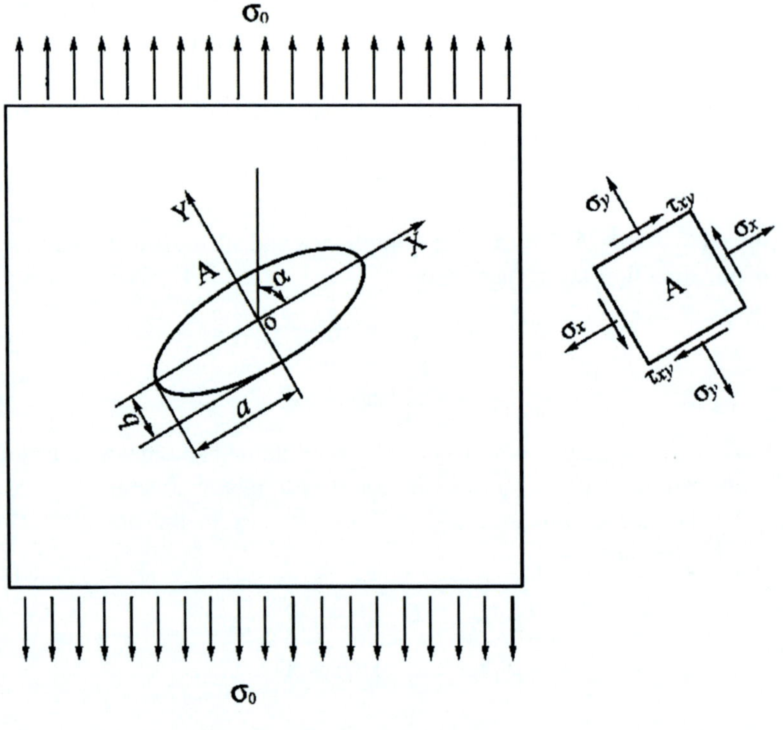

Fig. 6.19 Infinite plate containing an elliptic hole subjected to a nominal stress σ_0

$$\psi_0(z_2) = -\frac{i\sigma_0(a - is_2b)}{4(s_1 - s_2)}\left\{\frac{b(s_1 \sin 2\alpha + 2\cos^2\alpha)}{z_2 + \sqrt{z_2^2 - (a^2 + s_2^2b^2)}} + \frac{ia(2s_1 \sin^2\alpha + \sin 2\alpha)}{z_2 + \sqrt{z_2^2 - (a^2 + s_2^2b^2)}}\right\} \tag{6.85}$$

and the coordinate transformation is:

$$z_1 = x + s_1y = x_1 + iy_1 \tag{6.86}$$

$$z_2 = x + s_2y = x_2 + iy_2 \tag{6.87}$$

s_i is the root of the biharmonic characteristic equation of anisotropic material as follows:

$$a_{11}s^4 - 2a_{16}s^3 + (2a_{12} + a_{66})s^2 - 2a_{26}s + a_{22} = 0 \tag{6.88}$$

where a_{ij} is the parameter of the biharmonic characteristic equation for anisotropic material and s_i is the root of the biharmonic characteristic equation for anisotropic material:

$$s_1 = \alpha_1 + i\beta_1 \tag{6.89}$$

$$s_2 = \alpha_2 + i\beta_2 \tag{6.90}$$

$$s_3 = \alpha_1 - i\beta_1 \tag{6.91}$$

$$s_4 = \alpha_2 - i\beta_2 \tag{6.92}$$

Thus, the transformation of Eqs. 6.86, 6.87 and 6.89–6.92 also gives $x_1 = x + \alpha_1 y$, $y_1 = \beta_1 y$ and $x_2 = x + \alpha_2 y$, $y_2 = \beta_2 y$.

For orthotropic materials, $a_{16} = a_{26} = 0$, $a_{11} = \frac{1}{E_x}$, $a_{22} = \frac{1}{E_y}$, $a_{12} = -\frac{\nu_x}{E_x}$, $a_{66} = \frac{1}{G_{xy}}$. Equation 6.88 then becomes:

$$a_{11}s^4 + (2a_{12} + a_{66})s^2 + a_{22} = 0 \tag{6.93}$$

Equation 6.93 only has imaginary roots; α_1, α_2, α_3, α_4 are the real parts of these roots; the imaginary parts $\beta_1 > 0$, $\beta_2 > 0$, $\beta_1 \neq \beta_2$. When the stress direction is along the OX axis, namely $\alpha = 0$, Eqs. 6.84 and 6.85 become:

$$\begin{aligned}\phi_0(z_1) &= \frac{i\sigma_0 b}{2(s_1 - s_2)} \times \frac{a - is_1b}{z_1 + \sqrt{z_1^2 - (a^2 + s_1^2b^2)}}\\ &= \frac{i}{2(s_1 - s_2)} \cdot \frac{\sigma_0 b}{a + is_1b}\left[1 - \frac{z_1}{\sqrt{z_1^2 - (a^2 + s_1^2b^2)}}\right]\end{aligned} \tag{6.94}$$

$$\psi_0(z_2) = -\frac{i\sigma_0 b}{2(s_1 - s_2)} \times \frac{a - is_1 b}{z_2 + \sqrt{z_2^2 - (a^2 + s_2^2 b^2)}}$$
$$= -\frac{i}{2(s_1 - s_2)} \cdot \frac{\sigma_0 b}{a + is_2 b}\left[1 - \frac{z_2}{\sqrt{z_2^2 - (a^2 + s_2^2 b^2)}}\right] \tag{6.95}$$

Usually, for most materials, $s_1 = i\beta_1$, $s_2 = i\beta_2$. Therefore Eqs. 6.81–6.83 give the stress on the cross section of $x = 0$ as:

$$\sigma_x = -\sigma_0 - \frac{\sigma_0 b}{\beta_1 - \beta_2}\left\{\frac{\beta_1^2}{a - \beta_1 b}\left(1 - \frac{\beta_1 y}{\sqrt{a^2 + \beta_1^2(y^2 - b^2)}}\right) - \frac{\beta_2^2}{a - \beta_2 b}\left(1 - \frac{\beta_2 y}{\sqrt{a^2 + \beta_2^2(y^2 - b^2)}}\right)\right\} \tag{6.96}$$

$$\sigma_y = \frac{\sigma_0 b}{\beta_1 - \beta_2}\left\{\frac{1}{a - \beta_1 b}\left(1 - \frac{\beta_1 y}{\sqrt{a^2 + \beta_1^2(y^2 - b^2)}}\right) - \frac{1}{a - \beta_2 b}\left(1 - \frac{\beta_2 y}{\sqrt{a^2 + \beta_2^2(y^2 - b^2)}}\right)\right\} \tag{6.97}$$

$$\tau_{xy} = 0 \tag{6.98}$$

According to Eqs. 6.96–6.98, the stress at $y = b$ on the cross section of $x = 0$ is derived as:

$$\sigma_x = -\sigma_0\left[1 + (\beta_1 + \beta_2)\frac{b}{a}\right] \tag{6.99}$$

$$\sigma_y = \frac{\sigma_0 b}{\beta_1 - \beta_2}\left\{\frac{1}{a - \beta_1 b}\left(1 - \beta_1\frac{b}{a}\right) - \frac{1}{a - \beta_2 b}\left(1 - \beta_2\frac{b}{a}\right)\right\} \tag{6.100}$$

$$\tau_{xy} = 0 \tag{6.101}$$

In the case of orthotropic laminates containing a round hole $a = b$, and then the stress at $y = b$ on the cross section of $x = 0$ is:

$$\sigma_x = -\sigma_0(1 + \beta_1 + \beta_2) \tag{6.102}$$

$$\sigma_y = \tau_{xy} = 0 \tag{6.103}$$

If the notched laminates are of finite width, a correction coefficient of the effective stress along the notch [25] is given as follows:

$$k = \frac{2 + (1 - 2a/w)^3}{3(1 - 2a/w)} \tag{6.104}$$

Substituting the above into Eq. 6.102, the effective stress at $y = b$ on the cross section of $x = 0$ of the orthotropic laminates with finite width is obtained as:

$$\sigma_x = -k\sigma_0(1+\beta_1+\beta_2) = -\sigma_0(1+\beta_1+\beta_2)\left[\frac{2+(1-2a/w)^3}{3(1-2a/w)}\right] \tag{6.105}$$

$$\sigma_y = \tau_{xy} = 0 \tag{6.106}$$

6.7 Composite Damage Evolution Equation and Generalized σ–N Surface

Analogous to the J-integral or energy release rate G in fracture mechanics, the concept of a damage strain energy release rate is introduced and interpreted as the driving force and criterion for fatigue damage propagation. The damage strain energy release rate $\boldsymbol{\Delta\Gamma}$ is defined as:

$$\Delta\Gamma_i = \frac{\partial \Delta W_e(\sigma, D)}{\partial D_i} \tag{6.107}$$

Substituting Eq. 6.66 into Eq. 6.107 yields:

$$\Delta\Gamma_i = 2\sigma_a^T : \frac{\partial \boldsymbol{S}(\boldsymbol{D})}{\partial D_i} : \sigma_a \tag{6.108}$$

By substituting Eq. 6.74 into Eq. 6.107, the damage strain energy release rate for the orthotropic material in two-dimensional stress state is as:

$$\Delta\Gamma = \left\{\begin{matrix} \frac{2\sigma_{a1}^2}{E_1(1-D_1)^2} - \frac{\nu_{12}\sigma_{a1}\sigma_{a2}}{E_1\sqrt{(1-D_1)^3(1-D_2)}} - \frac{\nu_{21}\sigma_{a1}\sigma_{a2}}{E_2\sqrt{(1-D_1)^3(1-D_2)}} \\ \frac{2\sigma_{a2}^2}{E_2(1-D_2)^2} - \frac{\nu_{12}\sigma_{a1}\sigma_{a2}}{E_1\sqrt{(1-D_1)(1-D_2)^3}} - \frac{\nu_{21}\sigma_{a1}\sigma_{a2}}{E_2\sqrt{(1-D_1)(1-D_2)^3}} \\ \frac{2\tau_{a12}^2}{G_{12}(1-D_6)^2} \end{matrix}\right\} \tag{6.109}$$

Similarly, the three-dimensional damage evolution equation ($\frac{\mathrm{d}\boldsymbol{D}}{\mathrm{d}N^*} - \Delta\Gamma$ curve), analogous to the fatigue crack growth rate equation in fracture mechanics, can take the following power function form as:

$$\frac{dD_i}{dN^*} = C_i(\Delta\Gamma_i)^{m_i}(1-R_i)^{q_i} = C_i\left(2\sigma_a^T : \frac{\partial \boldsymbol{S}(\boldsymbol{D})}{\partial D_i} : \sigma_a\right)^{m_i}(1-R_i)^{q_i} \tag{6.110}$$

where D_i is the damage variable. $C_i m_i$ and q_i are the material constants of composites fatigue $\mathbf{dD}/\mathrm{d}N^* - \boldsymbol{\Delta\Gamma}$ curve. R is the stress ratio. $i = 1, 2, \ldots, 6$. In the case of a one-dimensional stress level, Eq. 6.110 degenerates into:

$$\frac{dD_i}{dN^*} = C_i(\Delta\Gamma_i)^{m_i} = C_i\left(2\sigma_a^T : \frac{\partial \boldsymbol{S}(\boldsymbol{D})}{\partial D_i} : \sigma_a\right)^{m_i} \tag{6.111}$$

By substituting Eq. 6.109 into Eq. 6.110, the damage evolution equation of the orthotropic material in two-dimensional stress state is obtained:

$$\begin{Bmatrix} \frac{dD_1}{dN^*} \\ \frac{dD_2}{dN^*} \\ \frac{dD_3}{dN^*} \end{Bmatrix} = \begin{Bmatrix} C_1\left[\frac{2\sigma_{a1}^2}{E_1(1-D_1)^2} - \frac{\nu_{12}\sigma_{a1}\sigma_{a2}}{E_1\sqrt{(1-D_1)^3(1-D_2)}} - \frac{\nu_{21}\sigma_{a1}\sigma_{a2}}{E_2\sqrt{(1-D_1)^3(1-D_2)}}\right]^{m_1}(1-R_1)^{q_1} \\ C_2\left[\frac{2\sigma_{a2}^2}{E_2(1-D_2)^2} - \frac{\nu_{12}\sigma_{a1}\sigma_{a2}}{E_1\sqrt{(1-D_1)(1-D_2)^3}} - \frac{\nu_{21}\sigma_{a1}\sigma_{a2}}{E_2\sqrt{(1-D_1)(1-D_2)^3}}\right]^{m_2}(1-R_2)^{q_2} \\ C_3\left[\frac{2\tau_{a12}^2}{G_{12}(1-D_6)^2}\right]^{m_3}(1-R_3)^{q_3} \end{Bmatrix} \tag{6.112}$$

Thus, from Eq. 6.112, the damage evolution equation for a transversely isotropic material in a two-dimensional stress state is:

$$\begin{Bmatrix} \frac{dD_1}{dN^*} \\ \frac{dD_2}{dN^*} \\ \frac{dD_3}{dN^*} \end{Bmatrix} = \begin{Bmatrix} C_1\left[\frac{2\sigma_{a1}^2}{E_1(1-D_1)^2} - \frac{2\nu_{12}\sigma_{a1}\sigma_{a2}}{E_1\sqrt{(1-D_1)^3(1-D_2)}}\right]^{m_1}(1-R_1)^{q_1} \\ C_2\left[\frac{2\sigma_{a2}^2}{E_1(1-D_2)^2} - \frac{2\nu_{12}\sigma_{a1}\sigma_{a2}}{E_1\sqrt{(1-D_1)(1-D_2)^3}}\right]^{m_2}(1-R_2)^{q_2} \\ C_3\left[\frac{\tau_{a12}^2}{(1-D_6)^2}\left(\frac{1}{E_1}+\frac{\nu_{12}}{E_1}\right)\right]^{m_3}(1-R_3)^{q_3} \end{Bmatrix} \tag{6.113}$$

In the case of the consideration of the one-dimensional stress level, Eq. 6.113 becomes:

$$\begin{Bmatrix} \frac{dD_1}{dN^*} \\ \frac{dD_2}{dN^*} \\ \frac{dD_3}{dN^*} \end{Bmatrix} = \begin{Bmatrix} C_1\left[\frac{2\sigma_{a1}^2}{E_1(1-D_1)^2} - \frac{2\nu_{12}\sigma_{a1}\sigma_{a2}}{E_1\sqrt{(1-D_1)^3(1-D_2)}}\right]^{m_1} \\ C_2\left[\frac{2\sigma_{a2}^2}{E_1(1-D_2)^2} - \frac{2\nu_{12}\sigma_{a1}\sigma_{a2}}{E_1\sqrt{(1-D_1)(1-D_2)^3}}\right]^{m_2} \\ C_3\left[\frac{\tau_{a12}^2}{(1-D_6)^2}\left(\frac{1}{E_1}+\frac{\nu_{12}}{E_1}\right)\right]^{m_3} \end{Bmatrix} \tag{6.114}$$

From Eq. 6.113, the damage evolution equation for transversely isotropic laminates of finite width containing a round hole the major direction of which coincides with the coordinate axis, under uniaxial tensile or compression loading of $\sigma_1 = \sigma$ and with $\sigma_2 = \tau_{12} = 0$, is:

$$\frac{dD}{dN^*} = C\left\{\frac{2\sigma_a^2(1+\beta_1+\beta_2)^2}{E(1-D)^2}\left[\frac{2+(1-2a/w)^3}{3(1-2a/w)}\right]^2\right\}^m\left(\frac{2\sigma_a}{\sigma_m+\sigma_a}\right)^q \tag{6.115}$$

The integral form of the equation above is:

$$N^* = \frac{\left[18E(1-2a/w)^2\right]^m\left[(1-D_0)^{2m+1}-(1-D_{cr})^{2m+1}\right]}{C\sigma_a^{2m}(2m+1)(1+\beta_1+\beta_2)^{2m}\left[2+(1-2a/w)^3\right]^{2m}}\left(\frac{2\sigma_a}{\sigma_m+\sigma_a}\right)^{-q} \tag{6.116}$$

where D_0 is the damage formation value and D_{cr} is the critical damage value at failure.

Equation 6.116 is a $\sigma_a - \sigma_m - N^*$ surface describes the relation between the stress amplitude σ_a, mean stress σ_m and fatigue life N^*, in a three-dimensional coordinate system. In the case of a one-dimensional stress level, Eq. 6.116 becomes:

$$N^* = \frac{\left[18E(1-2a/w)^2\right]^m \left[(1-D_0)^{2m+1} - (1-D_{cr})^{2m+1}\right]}{C\sigma_a^{2m}(2m+1)(1+\beta_1+\beta_2)^{2m}\left[2+(1-2a/w)^3\right]^{2m}} \tag{6.117}$$

Equation 6.117 is a σ–N^* curve in the plane coordinate system expressing the relation of the stress σ_a amplitude and fatigue life N^*.

References

1. Reifsnider KL (1990) Fatigue of composite materials. Elsevier, Oxford
2. Talreja R (2000) Fatigue of polymer matrix composites. In: Talreja R, Manson JE, Kelly A, Zweben C (eds) Comprehensive composite materials (vol 2): polymer matrix composites. Elsevier, Oxford
3. Read PJCL, Shenoi RA (1995) A review of fatigue damage modelling in context of marine FRP laminates. Mar Struct 8(1):257–278
4. Bader MG (1991) Composites. Butterworth Heinemann, Oxford
5. Soutis C, Smith EC, Matthews FL (2000) Predicting the compressive engineering performance of carbon fibre-reinforced plastics. Compos Part A 31:531–536
6. Wang CM, Shin CS (2002) Residual properties of notched [0/90]45AS4/PEEK composite laminates after fatigue and re-consolidation. Compos Part B 33:67–76
7. Nuismer RJ, Whitney JM (1975) Uniaxial failure of composites laminates containing stress concentrations. ASTM STP 593, pp 117–142
8. Xiong JJ, Shenoi RA, Wang SP, Wang WB (2004) On static and fatigue strength determination of carbon fibre/epoxy composites. Part I: experiments. J Strain Anal Eng Des 39(5):529–540
9. Beheshty MH, Harris B, Adam T (1993) An empirical fatigue-life model for high-performance fibre composites with and without impact damage. Compos Part A Appl Sci Manuf 30(8):971–987
10. Kaminski M (2002) On probabilistic fatigue models for composite material. Int J Fatigue 24:477–495
11. Yang JN, Liu MD (1977) Residual strength degradation model and theory of periodic proof tests for Graphite/Epoxy laminates. J Compos Mater 11:176–203
12. Xiong JJ, Shenoi RA (2004) Two new practical models for estimating reliability-based fatigue strength of composites. J Compos Mater 38(14):1187–1209
13. Xiong JJ, Shenoi RA, Wang SP, Wang WB (2004) On static and fatigue strength determination of carbon fibre/epoxy composites. Part II: theoretical formulation. J Strain Anal Eng Des 39(5):541–548
14. Xiong JJ, Li YY, Zeng BY (2008) A strain-based residual strength model of carbon fibre/epoxy composites based on CAI and fatigue residual strength concepts. Compos Struct 85:29–42
15. Jessen SM, Plumtree A (1991) Continuum damage mechanics applied to cyclic behavior of glass fibre composites pultrusion. Composites 22:181–190

16. Shen W, Peng L, Yang F, Shen Z (1987) Generalized elastic damage theory and its application to composite plate. Eng Fract Mech 28:403–412
17. Chow CL, Wang J (1987) An anisotropic theory of elasticity for continuum damage mechanics. Int J Fract 33:3–16
18. Xiong JJ, Shenoi RA (2004) A two-stage theory on fatigue damage and life prediction of composites. Compos Sci Technol 64(9):1331–1343
19. Phillips HJ, Shenoi RA (1998) Damage tolerance of laminated tee joints in FRP ship structures. Composites 29A:465–478
20. Phillips HJ, Shenoi RA (1999) Damage mechanics of top-hat stiffeners used in FRP ship construction. Mar Struct 12:1–19
21. Hwang W, Han KS (1986) Cumulative damage models and multi-stress fatigue life prediction. J Compos Mater 20:125–153
22. Weibull W (1961) Fatigue testing and analysis of results. Macmillan Company, New York
23. Sidoroff F (1981) Description of anisotropic damage application to elasticity: physical non-linearities in structure analysis. In: IUTAM Symposium. Springer-Verlag, Berlin, pp 237–244
24. Lekhnitskii SG (1968) Anisotropic plates. Gordan and Breach Science Publishers, New York
25. Whitney JM, Kim RY (1977) Effect of stacking sequence on the notched strength of laminated composites. ASTM STP 617:229–242

Chapter 7
Chaotic Fatigue

7.1 Introduction

Because the stochastic nature of fatigue induced damage in macro-scale results is derived from the random behaviour at microscopic and atomic levels, and in order to explain the macro stochastic phenomena and reveal the failure mechanisms, the nanoscopic or microscopic-structural fracture process analysis of solids has become a popular and interesting research topic. Actually, fatigue crack formation and propagation in a solid, resulting from the breaking of atomic bonds, is a highly non-linear dynamic process, particularly in the vicinity of a crack tip. Many researchers [1–9] have reported on the fractal nature of fracture surfaces and the existence of chaotic processes occurring in vicinity of a crack tip: the existence of chaotic processes was consistent with the measured fractal dimension of the fracture surface. One might therefore anticipate that the chaotic processes would somehow contribute to fatigue crack initiation and propagation. Thus, Markworth et al. [10], Mohan et al. [11] presented a simple atomistic model of a crack tip which consists of a linear array of four atoms, with nearest neighbours interacting via a Morse-function potential (shown in Fig. 7.1), obtained the determining conditions under which the two inner atoms, which represent the crack tip, oscillate in a chaotic or aperiodic manner in response to the driving and damping—see Eq. 7.1. Using this simple atomistic model, chaotic behaviour is found to be associated with larger driving amplitudes and smaller damping constants; extreme sensitivity to initial conditions in the long-term behaviour is demonstrated qualitatively by examining the effect of minute changes in initial conditions on the dynamics of motion as well as by calculation of Lyapunov exponents.

$$m\ddot{s}_i = -c\dot{s}_i - v'(r_i;k) + v'(R - r_1 - r_2;k) \tag{7.1}$$

with

$$r_i = r_0 + s_i - u_i$$

J. J. Xiong and R. A. Shenoi, *Fatigue and Fracture Reliability Engineering*,
Springer Series in Reliability Engineering, DOI: 10.1007/978-0-85729-218-6_7,

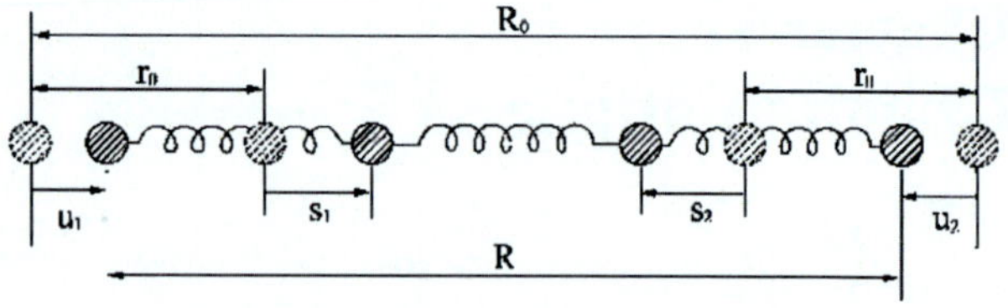

Fig. 7.1 An illustration of the crack-tip model

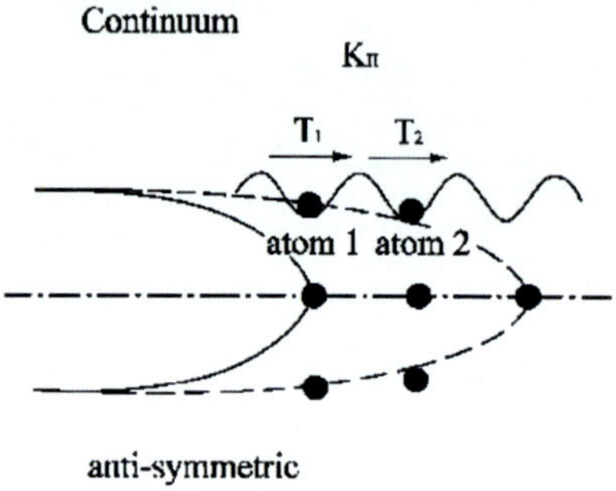

Fig. 7.2 Dislocation emission model

$$R = R_0 - u_1 - u_2$$

$$v(r; k) = kD\{\exp[-2\alpha(r - r_e)] - 2\exp[-\alpha(r - r_e)]\}$$

where for the two inner atoms, $i = 1, 2$; m is the atomic mass and c is the damping constant. R_0 is the overall length, at which the chain is being held if the two end atoms were not oscillating. r_0 is the distance from an outer atom to its adjacent inner atom when the potential energy of the entire chain has a local minimum value corresponding to the given value of R_0. The dot represents differentiation with respect to time, and the prime denotes differentiation with respect to the first argument of the potential function. $v(r; k)$ is the potential energy associated with the bond at interatomic distance r, i.e. the well known Morse function. kD is the depth of the potential well at the separation r_e, which corresponds to the minimum potential energy, and α is a measure of the 'hardness' of the potential. The parameter k is a positive, dimensionless quantity which is set equal to one for the two outer bonds. Tan and Yang [12–14] established several atom-string models for a crack tip and formulated its interaction with the overlapping cracked continuum a model (shown in Fig. 7.2 and Eq. 7.2) for dislocation emission under the fracture modes 1 and 2 loading to describe an atomistic sliding strip ahead of a crack and related the continuum mechanics field with the crack tip atomic motion. It was found that the atom-string system changes from deterministic to chaotic motion with the remote load increase.

$$m\ddot{u}_i = T_i(t)a_0^2 + \tau(u_i)a_0^2 - f(a_0 + u_2 - u_1) \tag{7.2}$$

where $i = 1, 2$, m is the atomic mass and a_0 is the equilibrium distance of a pair of atoms. $T_1(t)$ and $T_2(t)$ are the transverse confining stresses on atom 1 and atom 2 respectively, caused by the surrounding continuum. $f(a_0 + u_2 - u_1)$ is the interatomic force function of a pair of atoms. Due to the dissipative nature of fatigue damage characterization, only the most decisive parameters defining the fatigue damage process should be considered to establish the governing kinetic equation for a fatigue damage system. To consider the influence of the parameters omitted, an appropriate stochastic analysis for the governing kinetic equation has to be taken into account. Xiong et al. [15, 16] established a motion model of a pair of atoms at a crack tip for the fracture mode I and a stochastic governing kinetic equation. Using the singular point theory of a one-dimensional diffusion process and the stochastic averaging approach of the energy envelope (to determine exact stationary solutions of stochastically excited and dissipated integrable and partially integrable Hamilton systems), a micro-model to describe the atomic movement at the crack tip in a homoclinic bifurcation fatigue damage system (which often occurs when a periodic orbit collides with a saddle point), which is in the presence of stochastic perturbation, is established. After the study on the characteristic of the diffusion exponent, the drift exponent and the character exponent of the fatigue damage diffusion process on a singular boundary, the bifurcation behavior of a homoclinic bifurcation fatigue damage system, which is in the presence of parametric white noise, is examined.

From the earlier works, it is clear that most of the nonlinear analysis on the micro-crack tip is based on numerical simulation approaches. Hence, the purpose of this chapter is to examine atomic-scale chaotic and stochastic bifurcation process in order to estimate their influence on the fatigue fracture process. Towards this end, a model describing atomic motion at a crack tip is proposed to more closely approximate the actual crack tip conditions by consideration of the external driving force, internal atomic interaction caused by Leonard-Jones potential energy and nonlinear damping force resulting from the surroundings in the vicinity of the crack tip. This model reveals the highly nonlinear kinetic property of the dislocation emission process. Particular attention is paid to the effects of the driving and damping process on the dynamics at the crack tip and on the occurrence and extent of chaotic dynamics. The long-term behaviour and disregarded transient chaos of the crack-tip atomic motions are predicted. The chaotic motion occurrence of dislocation nucleation is also forecast. By using the randomization of deterministic fatigue damage equation, the stochastic differential equation and the Fokker–Planck equation of fatigue damage affected by random fluctuations are derived, whose solutions provide the probability distributions of fatigue crack formation and propagation.

7.2 Nonlinear Differential Kinetic Model of Atomic Motion at Crack Tip

As shown in Fig. 7.3, a symmetrical crack is viewed as a semi-infinite slit in this paper, in an otherwise unbounded solid, with loading applied at infinity through a

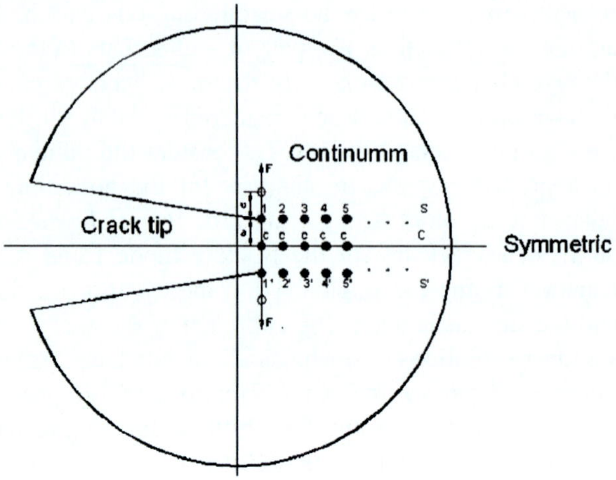

Fig. 7.3 Crack tip configuration [12]

crack tip singularity field, characterized by a remote action force F. Therefore, the motion of the atoms at the crack tip is symmetric to the central line; that is, the atoms c_1, c_2, ... remain immobile during the process of dislocation emission and the motion of atoms 1, 2, ... is considered. This model is assumed to satisfy the following four conditions.

(1) The material in a very small region around this crack tip consists of several interesting triples of atoms represented by $1c_1 1', 2c_2 2', \ldots$, respectively, i.e. these triples of atoms ahead of the crack tip represent the crack tip to realistically connect and bond the cleavage continuums at the upper and lower crack surfaces.
(2) Rice [17] discussed the equilibrium (or static) process of dislocation emission from a crack tip based on the Peierls concept. A slip plane with periodic shear stress and sliding displacement relation is embedded in an elastic continuum to incorporate an atomistic description of the dislocation core (shown in Fig. 7.4). Rice's static analysis introduced a new solid state parameter γ_{us}, termed as the unstable stacking energy. He also established the Irwin energy release rate G as:

$$G = \Phi(\delta_{tip}) \tag{7.3}$$

where $\delta = u^+ - u^-$, δ_{tip} is the opening displacement δ ahead of the crack tip. At $G = \gamma_{us}$, the incipient dislocation configuration loses stability to become the nucleation of a full dislocation.
From Eq. 7.3, it is noteworthy that the Irwin energy release rate G is a function of the opening displacement δ ahead of the crack tip, or the relative distance between the Atoms 1 and $1'$ (shown in Fig. 7.3). Consequently, the motion of the atom 1 along the r direction plays an important role in the fracture process of actual material.

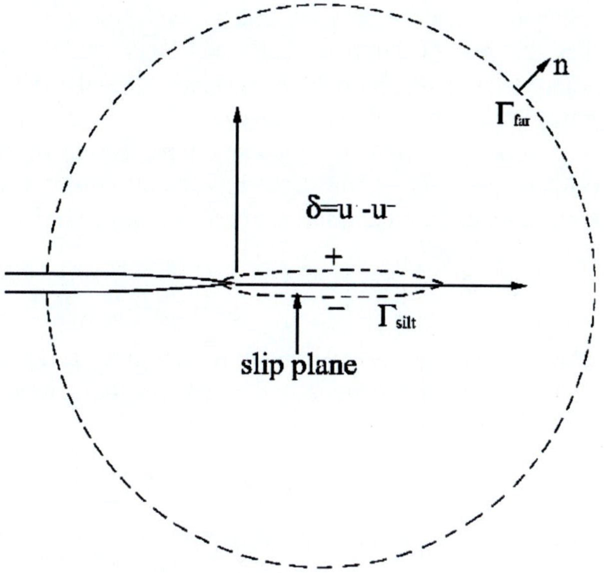

Fig. 7.4 Coincident crack and slip plane [17]

(3) The interaction of the nearest neighbours of the crack-tip atoms 1 and c_1 along the r direction is determined through the Leonard-Jones interatomic potential energy; the interaction force between atom 1 and other neighbours is regarded as the damping force.
(4) The slip planes are parallel to the crack extension plane, and both the spacing between slip planes and the length of the Burgers vector are a_0.

The classical formulation for interatomic potentials assumes that all particles in a crystal are bonded by the central interaction between a pair of atoms. This approximation leads to an effective quantification of the interatomic actions. The conventional Leonard-Jones potential can be transformed to a solid, resulting in the following form for the interatomic potential energy [18–20]:

$$\frac{e(r)}{e_0} = -3\frac{a_0^6}{r^6} + 2\frac{a_0^9}{r^9} \tag{7.4}$$

From Eq. 7.4, Leonard-Jones interatomic potential energy $e(r)$ has a similar formulation to the principle of universal gravitation; if $r \to \infty$, then $e(r) = 0$ and if $r \to 0$, then $e(r) \to \infty$.

The interatomic force f can be obtained as:

$$f(r) = -\frac{de(r)}{dr} = 18e_0{a_0}^6\left(\frac{a_0^3}{r^{10}} - \frac{1}{r^7}\right) \tag{7.5}$$

Suppose that each atom only interacts with the n_0 nearest neighboring atoms, and the potentials of all paired interactions are identical and independent from other interactions. At an equilibrium configuration free of external load, $e_0 = Ca_0^3/3n_0$. C is the bulk modulus for hydrostatic stressing; a_0 is the equilibrium distance of a pair of atoms; C and a_0 can be obtained from experiments. If $f(r) > 0$, then this pair of atoms exclude each other; if $f(r) < 0$, then they mutually inter-attract. Based on the balance distance a_0 and the balance energy e_0, it can be proved that $f(a_0) = 0$, $e(a_0) = -e_0$.

It is assumed that the damping force acting on atom 1 satisfies the viscous principle [10, 11], i.e. the damping force is proportional to and directed opposite to their velocity, the damping coefficient is a linear function of the displacement of the atom. That is:

$$f_d = -(ku + c)\dot{u}$$

where k and c are two constants, which may be determined by experiments.

The governing equation of crack tip atom motion, from Newton's second principle, is then:

$$m\ddot{u} = F - (ku + c)\dot{u} + f(r) \tag{7.6}$$

where r represents the displacement of crack tip atom 1 or 1' to the central atom c_1, $r = a_0 + u$.

Substitution of Eq. 7.5 into Eq. 7.6 gives the governing equation of atomic motion as:

$$m\ddot{u} = F - (ku + c)\dot{u} - \frac{18e_0}{a_0\left(1 + \frac{u}{a_0}\right)^7} + \frac{18e_0}{a_0\left(1 + \frac{u}{a_0}\right)^{10}} \tag{7.7}$$

Because $u \ll a_0$, therefore u/a_0 is a small quantity. Using Taylor's expansion to express the last two items of Eq. 7.7 and neglecting the higher-order forms, Eq. 7.7 becomes:

$$\ddot{u} = C_3u^3 + C_2u^2 + C_1u + P - (Ku + \delta)\dot{u} \tag{7.8}$$

where $C_1 = -54e_0a_0^{-2}/m$, $C_2 = 486e_0a_0^{-3}/m$, $C_3 = -2248e_0a_0^{-4}/m$, $P = F/m$, $K = k/m$, $\delta = c/m$.

Letting $v = \dot{u}$, from Eq. 7.8, one can gain:

$$\dot{v} = C_3u^3 + C_2u^2 + C_1u + P - (Ku + \delta)v$$

Thus, the equivalent system of Eq. 7.8 is:

$$\begin{cases} \dot{u} = v \\ \dot{v} = C_3u^3 + C_2u^2 + C_1u + P - (Ku + \delta)v \end{cases} \tag{7.9}$$

As is well known, the external driving force P applied on the atoms at a crack tip cannot be kept constant but is accompanied by a fluctuation varying minutely with time t. Therefore, P can be viewed as a perturbance $\varepsilon\gamma\cos\omega_0 t$ with frequency ω_f and large amplitude γ dependent on time t. Furthermore, in the case of weak damping, the internal damping force is a small quantity and can be expressed as $\varepsilon\delta v$. Thus Eq. 7.9 becomes:

$$\begin{aligned} \dot{u} &= v \\ \dot{v} &= C_3u^3 + C_2u^2 + C_1u + \varepsilon(\gamma \cos \omega_0 t - \delta v) \end{aligned} \tag{7.10}$$

where ε is the small quantity.

It should be emphasized that P in Eq. 7.9 represents the most decisive influence of the external conditions on this pair of atoms. Considering the influence of the omitted parameters, and applying a random perturbance on the kinetic system as given in Eq. 7.9, then P is randomized into a stochastic driving oscillation of a unit white noise process and Eq. 7.9 becomes:

$$\begin{aligned} \dot{u} &= v \\ \dot{v} &= C_3u^3 + C_2u^2 + C_1u - K(u + x_0)v + P_0vX(t) \end{aligned} \tag{7.11}$$

where $x_0 = \delta/K$.

From Eq. 7.6, it is clear that if $k = 0$, then Eq. 7.6 can become the one-dimensional form of Eq. 7.1; if $k = 0$ and $c = 0$, then Eq. 7.6 can be the one-dimensional form of Eq. 7.2. Therefore, it is concluded that under some conditions, Eq. 7.6 is consistent with Eqs. 7.1 and 7.2, and can be expected to reveal and verify that the atomic motion behaviour can be chaotic even when the dissipation or damping of energy is found or proved by using a series of qualitative and quantitative checks. As mentioned earlier [11], because of their chaotic nature, atomic motion at a crack-tip may be extremely difficult, even impossible, to observe, due to the size scale involved. Nevertheless, the chaotic atomic motion predictions obtained using the models of Eq. 7.6 do seem to provide as useful insight into the fracture processes of actual materials as Eqs. 7.1 and 7.2.

Using the analogy of Eq. 7.1, in the models expressed in Eq. 7.6, it is particularly emphasized that the energy exchange between the crack tip and its surroundings is simulated by an external remote driving force, that is, forcing a pair of atoms at the crack-tip to oscillate sinusoidally with time, and by allowing the motion of these two atoms to be nonlinearly damped. This energy exchange included driving process simulates thermally activated oscillations of crack-tip atoms as well as loss of energy by damping processes. The damping may, at least qualitatively, be related to dissipative mechanisms that may occur at the tip of a crack. By means of the same approximate potential and a pair of atoms in Eq. 7.2 to represent the non-linear nature of the interatomic bond and the crack-tip in real fracture solid, the simple atomic model of a crack tip expressed in Eq. 7.6 is used to describe the atomic motion in the highly non-linear crack-tip region and to illustrate the extent to which chaos can play a role of crucial importance in its dynamic behaviour.

As mentioned above, Eq. 7.6 has been extended to the Hopf bifurcation dynamical Eq. 7.9, chaotic dynamical Eq. 7.10 and stochastic bifurcation Eq. 7.11, which can be used to make long-term predictions regarding bifurcation and chaotic behavior and to demonstrate the extreme sensitivity to initial external driving conditions. These extensions of Eq. 7.6 can simulate the atomic-motion processes at the real tip of a crack and to evaluate their influence on the fracture process. Of course, the models expressed in Eqs. 7.9–7.11 are not as suitable for the simulation of crack propagation details in a real three-dimensional fracture solid as other approaches based on techniques of molecular dynamics [21–23]. However, they offer the opportunity to examine, in detail, the extremely complex, highly non-linear behaviour of a single pair of mutually interacting crack-tip atoms and to relate this behaviour to the overall fracture process. In short, what is evident from most of the above discussion is that the atomic models expressed in Eqs. 7.6, 7.9–7.11 are different from those in Eqs. 7.1 and 7.2, while Eq. 7.6 has combined some characteristics of Eqs. 7.1 and 7.2.

7.3 Hopf Bifurcation of Atomic Motion at Crack Tip

In general, the chaotic behaviour of nonlinear dynamical system can be explored based on two classes of methodology: (1) numerical simulation of phase space orbit [24], which includes three methods of the Poincare Map, Lyapunov exponents and fractal dimension; (2) analytic differentiable manifold of vector field [25], which contains three approaches of the Hopf bifurcation (local bifurcation), global bifurcation (chaos) and stochastic bifurcation. The applied condition of the numerical simulation is that the parameters of the nonlinear dynamical system should be given or assumed, and then the Poincare Map, Lyapunov exponents and fractal dimension can be simulated, while the analytic differentiable manifold can be used to determine the analytic expression of the system parameters under the occurrence of the Hopf bifurcation, global bifurcation and stochastic bifurcation. So it is obvious that the analytic solutions of nonlinear dynamical system can be easily obtained through the analytic differentiable manifold method when the chaotic motion occurs.

By means of the analytic differentiable manifold theory [25], Hopf bifurcation describes the limited mutations of the system behaviour, whereas the global bifurcation depicts the infinite mutations. The limited mutations are governed by Hopf bifurcation parameters, while the infinite mutations are analyzed based on the classical Melnikov analysis for homoclinic bifurcations. As a result, three equilibrium points of Eq. 7.9 are as follows $P_1(x_1, 0)$, $P_2(x_2, 0)$, $P_3(x_3, 0)$, with:

$$x_1 = \sqrt[3]{-\frac{q}{2}+\sqrt{\left(\frac{q}{2}\right)^2+\left(\frac{p}{2}\right)^3}} + \sqrt[3]{-\frac{q}{2}-\sqrt{\left(\frac{q}{2}\right)^2+\left(\frac{p}{2}\right)^3}} - \frac{C_2}{3C_3} \tag{7.12}$$

$$x_2 = \omega_1\sqrt[3]{-\frac{q}{2}+\sqrt{\left(\frac{q}{2}\right)^2+\left(\frac{p}{2}\right)^3}} + \omega_1^2\sqrt[3]{-\frac{q}{2}-\sqrt{\left(\frac{q}{2}\right)^2+\left(\frac{p}{2}\right)^3}} - \frac{C_2}{3C_3} \tag{7.13}$$

$$x_3 = \omega_1^2\sqrt[3]{-\frac{q}{2}+\sqrt{\left(\frac{q}{2}\right)^2+\left(\frac{p}{2}\right)^3}} + \omega_1\sqrt[3]{-\frac{q}{2}-\sqrt{\left(\frac{q}{2}\right)^2+\left(\frac{p}{2}\right)^3}} - \frac{C_2}{3C_3} \tag{7.14}$$

where:

$$\omega_1 = \frac{-1 + i\sqrt{3}}{2}, \quad \omega_1^2 = \frac{-1 - i\sqrt{3}}{2}$$

$$p = \frac{3C_3C_1 - C_2^2}{3C_3^2} = -0.002059a_0^2$$

$$q = \frac{2C_2^3 - 27C_1C_2C_3}{27C_3^2} = -0.000578a_0^3$$

Substitution of the above parameters into Eqs. 7.12–7.14 yields:

$$x_1 = 0.1213a_0$$

$$x_2 = 0.0502a_0 + 0.0834a_0\omega - 0.0123a_0\omega^2$$

$$x_3 = 0.0502a_0 + 0.0834a_0\omega^2 - 0.0123a_0\omega$$

Taking the linear transformation of Eq. 7.9 as:

$$(u, v) = (\xi + x_1, \eta)$$

Then:

$$\begin{cases} \dot{\xi} = \eta \\ \dot{\eta} = a_3\xi^3 + a_2\xi^2 + a_1\xi + P - K(\xi + x_1)\eta - \delta\eta \end{cases} \tag{7.15}$$

where:

$$a_3 = C_3$$

$$a_2 = 3C_3x_1 + C_2$$

$$a_1 = 3C_3x_1^2 + 2C_2x_1 + C_1$$

Equation 7.15 exists for an equilibrium point of $P_1(0, 0)$. It can be written in the linearized matrix form as:

$$\begin{aligned} \begin{pmatrix} \dot{\xi} \\ \dot{\eta} \end{pmatrix} &= A\begin{pmatrix} \xi \\ \eta \end{pmatrix} + \begin{pmatrix} s(\xi, & \eta) \\ g(\xi, & \eta) \end{pmatrix} \\ &= \begin{bmatrix} 0 & 1 \\ a_1 & 0 \end{bmatrix}\begin{pmatrix} \xi \\ \eta \end{pmatrix} + \begin{pmatrix} 0 \\ a_3\xi^3 + a_2\xi^2 + P - K(\xi + x_1)\eta - \delta\eta \end{pmatrix} \end{aligned} \tag{7.16}$$

The characteristic equation of this is:

$$\begin{vmatrix} \lambda & -1 \\ -a_1 & \lambda \end{vmatrix} = 0$$

The characteristic roots are resolved as:

$$\lambda = \pm\sqrt{a_1}$$

When $-\omega^2 = a_1 < 0$, the above characteristic roots are a couple of conjugate complex numbers, namely $\lambda_{1,2} = \pm\, i\omega$, and the characteristic vectors are respectively:

$$Z_1 = (1,\ -i\omega)'$$

$$Z_2 = (1,\ -i\omega)'$$

Letting:

$$Y_1 = \frac{Z_1 + Z_2}{2} = (2,\ 0)'$$

$$Y_2 = \frac{Z_1 - Z_2}{2i} = (0,\ 2\omega)'$$

and:

$$B = (\, Y_2 \quad Y_1 \,) = \begin{pmatrix} 0 & 2 \\ 2\omega & 0 \end{pmatrix}$$

Then:

$$B^{-1} = \begin{pmatrix} 0 & -\frac{1}{2\omega} \\ -\frac{1}{2} & 0 \end{pmatrix}$$

Taking the linear transformation as:

$$\begin{pmatrix} \xi \\ \eta \end{pmatrix} = \begin{pmatrix} 2y \\ 2\omega x \end{pmatrix}$$

then the reduced canonical form of Eq. 7.16 is:

$$\begin{pmatrix} \dot{x} \\ \dot{y} \end{pmatrix} = B^{-1}AB\begin{pmatrix} x \\ y \end{pmatrix} + B^{-1}\begin{bmatrix} s(x,y) \\ g(x,y) \end{bmatrix} = \begin{pmatrix} 0 & -\omega \\ \omega & 0 \end{pmatrix}\begin{pmatrix} x \\ y \end{pmatrix} + \begin{pmatrix} S(x,y) \\ G(x,y) \end{pmatrix} \tag{7.17}$$

where:

$$\begin{aligned} S(x,y) &= -\frac{1}{\omega}\left(4a_3y^3 + 2a_2y^2 + a_1y - 2\omega Kxy - \omega Kxx_1 - \omega\delta x\right) \\ G(x,y) &= 0 \end{aligned} \tag{7.18}$$

According to the local bifurcation theory of differential dynamic systems [25], local bifurcation is identified through the Hopf bifurcation coefficient. This coefficient can be calculated as:

$$a = \frac{1}{16}(S_1 + S_2 + G_1 + G_2) + \frac{1}{16\omega}[S_3(S_4 + S_5) - G_3(G_4 + G_5) - S_4G_4 + S_5G_5] \tag{7.19}$$

with:

$$\begin{Bmatrix} S_1 \\ S_2 \\ S_3 \\ S_4 \\ S_5 \end{Bmatrix} = \begin{Bmatrix} \frac{\partial^3}{\partial x^3} \\ \frac{\partial^3}{\partial x \partial y^2} \\ \frac{\partial^2}{\partial x \partial y} \\ \frac{\partial^2}{\partial x^2} \\ \frac{\partial^2}{\partial y^2} \end{Bmatrix} S(x, y) \tag{7.20}$$

$$\begin{Bmatrix} G_1 \\ G_2 \\ G_3 \\ G_4 \\ G_5 \end{Bmatrix} = \begin{Bmatrix} \frac{\partial^3}{\partial x^2 \partial y} \\ \frac{\partial^3}{\partial y^3} \\ \frac{\partial^2}{\partial x \partial y} \\ \frac{\partial^2}{\partial x^2} \\ \frac{\partial^2}{\partial y^2} \end{Bmatrix} G(x, y) \tag{7.21}$$

Substituting Eqs. 7.17 and 7.18 into Eqs. 7.20 and 7.21 gives:

$$S_1 = 0,\ S_2 = 0,\ S_3 = 2K,\ S_4 = 0,\ S_5 = -\frac{4C_2}{\omega}$$

$$G_1 = 0,\ G_2 = 0,\ G_3 = 0,\ G_4 = 0,\ G_5 = 0$$

From Eq. 7.19 and the above calculation results, the Hopf bifurcation coefficient is obtained as:

$$a = -\frac{KC_2}{2\omega^2} \tag{7.22}$$

By means of the Hopf bifurcation criterion [25], if $a < 0$, then the differential kinetic system is weakly stable and supercritical bifurcation occurs (shown in Fig. 7.5); if $a > 0$, then the subcritical bifurcation happens in this system (shown in Fig. 7.6). From above analysis, it is apparent that no matter which case of bifurcation appears, the positions and velocities of atom 1 have multiple values due to the occurrence of the Hopf bifurcation. This implies that the Irwin energy release rate G calculated by Eq. 7.3 also has multiple values, which possibly leads to the occurrence of $G > \gamma_{us}$. Then the incipient dislocation configuration loses stability to become the nucleation of a full dislocation. The understanding of the relationship between the structure in phase space of the crack-tip kinetic system and the Irwin energy release rate G may, indeed, enable predictions of at least certain properties associated with apparently random behavior of real fracture process in macro-scale.

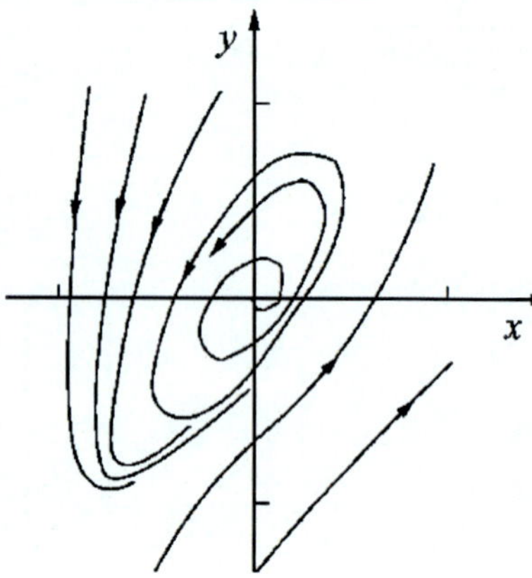

Fig. 7.5 Supercritical bifurcation

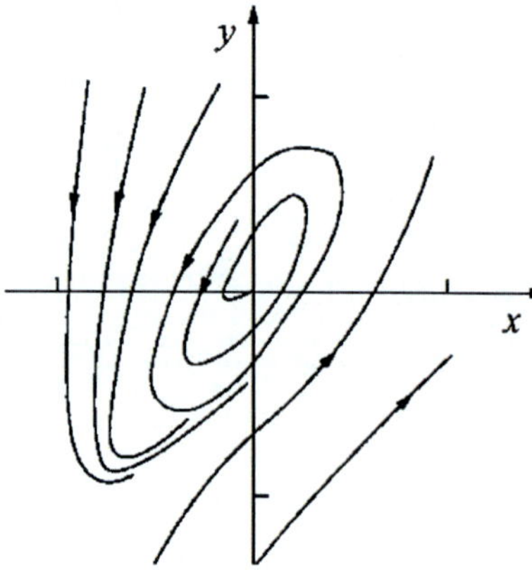

Fig. 7.6 Subcritical bifurcation

7.4 Global Bifurcation of Atomic Motion at Crack Tip

In case of that C_2 and ε are small enough to be neglected, then the topological structures of Eq. 7.10 in the phase-space have two forms as given by means of the differential manifold theory.

Form A: $H_1(u,v) = \frac{v^2}{2} - \frac{C_1}{2}u^2 - \frac{C_3}{4}u^4$

If $C_1 < 0$, $C_3 > 0$, and there exists an immovable point (0, 0) and nontrivial immovable points $\left(\pm\sqrt{-C_1/C_3}, 0\right)$, then the immovable point (0, 0) is steady, the nontrivial immovable points $\left(\pm\sqrt{-C_1/C_3}, 0\right)$ are two unsteady saddle points, and there is a heteroclinic trajectory to link these two saddle points in the limited zone $H_1 \in \left(0, \frac{C_1^2}{4C_3^2}\right)$.

Form B: $H_2(u,v) = \frac{v^2}{2} - \frac{a_3}{2}u^2 - \frac{a_1}{4}u^4$

If $C_1 > 0$, $C_3 < 0$, and there are three immovable points: (0, 0) and $\left(\pm\sqrt{-C_1/C_3}, 0\right)$, then the points$\left(\pm\sqrt{-C_1/C_3}, 0\right)$ are steady, (0, 0) is the unsteady saddle point, and a couple of symmetrical homoclinic trajectories with respect to the v axis of coordinates to link the saddle point (0, 0) in the limited zone $H_2 \in \left(-\frac{C_1^2}{4C_3^2}, 0\right)$. $H_2 = 0$ corresponds to the contour lines of these homoclinic trajectories.

For the topological structure form B, the unperturbed homoclinic trajectory of Eq. 7.10 through a point of $q_{\pm}^0(0) = (\pm\sqrt{-C_1/C_3}, 0)$ is:

$$q_+^0(t) = \left(\sqrt{-\frac{2C_1}{C_3}}\sec(ht), -\sqrt{-\frac{2C_1}{C_3}}\sec(ht)\cdot\tan(ht)\right) \tag{7.23}$$

$$q_-^0(t) = -q_+^0(t) \tag{7.24}$$

Based on the Melinkov theorem [25], Melinkov function of Eq. 7.10 can be obtained as:

$$\begin{aligned} M(t_0; \gamma, \delta, \omega_0) &= \int_{-\infty}^{+\infty} f\left[q^0(t)\right] \wedge g\left[q^0(t), t + t_0\right] dt \\ &= -\sqrt{-\frac{2C_1}{C_3}}\gamma \int_{-\infty}^{+\infty} \sec(ht)\cdot\tan(ht)\cdot\cos[\omega_0(t+t_0)]dt \\ &\quad + \frac{2C_1}{C_3}\delta \int_{-\infty}^{+\infty} \sec\left(h^2 t\right)\cdot\tan\left(h^2 t\right)dt \end{aligned}$$

$$M(t_0; \gamma, \delta, \omega_0) = \frac{4C_1}{3C_3}\delta - \sqrt{-\frac{2C_1}{C_3}}\gamma\pi\omega_0 \sec\left[h\left(\frac{\pi\omega_0}{2}\right)\right]\sin(\omega_0 t_0) \tag{7.25}$$

Defining

$$R^0(\omega_0) = \frac{4}{3\sqrt{2}\pi\omega_0}\sqrt{-\frac{C_1}{C_3}}\cos\left[h\left(\frac{\pi\omega_0}{2}\right)\right] \tag{7.26}$$

$R^0(\omega_0)$ follows the differential kinetic properties as below.

(1) If $R^0(\omega_0) < \gamma/\delta$, then $R^0(\omega_0)$ is the threshold value for the horse-shoe collar to appear. The steady manifold $W^s(p_\varepsilon)$ of the system in Eq. 7.10 intersects the unsteady manifold $W^u(p_\varepsilon)$, the chaos phenomenon appears, and here Poincare maps have two cross-cutting trajectories.
(2) If $R^0(\omega_0) > \gamma/\delta$, $W^s(p_\varepsilon) \cap W^u(p_\varepsilon) = \Phi$.
(3) $R^0(\omega_0) = \gamma/\delta$, then $M(t_0;\, \gamma,\, \delta,\, \omega_0)$ of Eq. 7.25 has the quadratic zeros, namely, in the γ–δ plane quadratic homoclinic tangencies of the steady manifold $W^s(p_\varepsilon)$ with the unsteady manifold $W^u(p_\varepsilon)$ occurs.

As mentioned above, the fact that chaotic vibrations could still occur under some conditions indicates that such behaviour may occur at a real crack tip and may be an important factor in random behaviour of crack-propagation using the analogy of Sect. 7.3.

7.5 Stochastic Bifurcation of Atomic Motion at Crack Tip

In order to discuss the stochastic bifurcation behaviour of the differential kinetic system of Eq. 7.11, according to Ito stochastic differential principle [26], the introduction of Hamilton function (energy envelope) H into Eq. 7.11 yields the Ito stochastic differential equations with regard to the variable u and the energy envelope H as follows:

$$du = \left(2H + \frac{C_3}{2}u^4 + \frac{2C_2}{3}u^3 + C_1u^2\right)^{\frac{1}{2}} dt \tag{7.27}$$

$$\begin{aligned} dH = {} & K\left(u + x_0 + \frac{\sigma^2}{2}\right)\left(2H + \frac{C_3}{2}u^4 + \frac{2C_2}{3}u^3 + C_1u^2\right)dt \\ & + K^{\frac{1}{2}}\sigma\left(2H + \frac{C_3}{2}u^4 + \frac{2C_2}{3}u^3 + C_1u^2\right)dW \end{aligned} \tag{7.28}$$

where $W(t)$ is the unit Wiener process and $\sigma = P_0K^{\frac{-1}{2}}$.

Since the energy envelope $H(t)$ shown in Eqs. 7.27 and 7.28 is the slowly varying stochastic process, according to Khasminskii limit theorem [27], it is clear that if $K \to 0$, then $H(t)$ weakly converges to a one-dimensional diffusion process in a probability form in the interval of $0 \le t \le T_0$ and $T_o \in o\left(\frac{1}{K^2}\right)$. Thus, the first approximation of $H(t)$ can be a one-dimensional diffusion process. The Ito stochastic differential equation governing this diffusion process can be obtained by the average of Eq. 7.28 with respect to u along the trajectory pertaining to $H(t)$ being constant. The average of Eq. 7.28 gives Ito stochastic differential equations as:

$$dH = \bar{m}(H)dt + \bar{\sigma}(H)dW \tag{7.29}$$

where $\bar{m}(H)$ is the drift coefficient; $\bar{\sigma}^2(H)$ is the diffusion coefficient.

$$\begin{aligned} \bar{m}(H) & = \left(\frac{1}{T(H)}\right)\int_{u^-}^{u^+}\left(u + x_0 + \frac{\sigma^2}{2}\right)\left(2H + \frac{C_3}{2}u^4 + C_1u^2\right)^{\frac{1}{2}} du \\ & = \left[\frac{1}{T(H)}\right][m_1(H) + m_2(H)]u \end{aligned} \tag{7.30}$$

$$\bar{\sigma}^2(H) = \left[\frac{\sigma^2}{T(H)}\right]\int_{u^-}^{u^+}\left(2H + \frac{C_3}{2}u^4 + C_1u^2\right)^{\frac{3}{2}} du = \left[\frac{\sigma^2}{T(H)}\right]\sigma^2(H) \tag{7.31}$$

$$m_1(H) = \int_{u^-}^{u^+}(u + x_0)\left(2H + \frac{C_3}{2}u^4 + C_1u^2\right)^{\frac{1}{2}} du$$

$$m_2(H) = \left(\frac{\sigma^2}{2}\right) \int_{u^-}^{u^+} \left(2H + \frac{C_3}{2}u^4 + C_1u^2\right)^{\frac{1}{2}} du$$

$$T(H) = \int_{u^-}^{u^+} \frac{1}{\sqrt{2H + \frac{C_3}{2}u^4 + C_1u^2}} du$$

u^+ and u^- are respectively the maximum and minimum roots of equation:

$$2H + \frac{C_3}{2}u^4 + C_1u^2 = 0 \tag{7.32}$$

For the topological structures in Form B in above section, one can get u^+ and u^- as:

$$u^{\pm} = \sqrt{-(C_1/C_3) \pm \sqrt{(C_1/C_3)^2 + 4H}} \tag{7.33}$$

and $C_1 > 0$, $C_3 < 0$, $H \in \left(-\frac{C_1^2}{4C_3^2}, 0\right)$. Therefore, by mathematical transformation, Eqs. 7.30 and 7.31 can be respectively derived as:

$$\begin{aligned} \bar{m}(H) = \frac{2}{15}\left[\frac{C_1}{C_3(2-n)}\right]^3 \Bigg\{ & \left(\frac{10x_0^2 + 5C_3\sigma^2}{C_1} - 2\right)(1-n)(2-n) \\ & - \left[\frac{10x_0^2 + 5C_3\sigma^2}{2C_1}(2-n)^2 - 4\left(n^2 - n + 1\right)\right]\frac{E_2(n)}{E_1(n)} \Bigg\} \end{aligned} \tag{7.34}$$

$$\bar{\sigma}^2(H) = \frac{4\sigma^2}{35}\left[\frac{C_1}{C_3(2-n)}\right]^4 \left[(1-n)^2(n+16) - 2(n-2)\left(n^2 + 4n - 4\right)\frac{E_2(n)}{E_1(n)}\right] \tag{7.35}$$

and:

$$H = \frac{C_1^2(n-1)}{C_3^2(2-n)^2} \tag{7.36}$$

While $K = 0$, the integrable Hamilton system of the topological structure (7.4) is:

$$\begin{aligned} \dot{u} &= v \\ \dot{v} &= C_3u^3 + C_1u \end{aligned} \tag{7.37}$$

The contour lines of the Hamilton function (7.4) are given by:

$$H(u, v) = \frac{v^2}{2} - \frac{C_1}{2}u^2 - \frac{C_3}{4}u^4 = 0 \tag{7.38}$$

$H(u, v)$ in Eq. 7.38 describes two close symmetrical homoclinic trajectories with respect to the v axis of coordinate linked with saddle-point (0, 0). From Eq. 7.38, it is obvious that if $H = 0$, then $n = 1$; if $H = -C_1^2/(4C_3^2)$, then $n = 0$.

From the solution of the one-dimensional Ito stochastic differential equations given in Eq. 7.29 for the case of $K \neq 0$, the sample properties of the one-dimensional energy envelope diffusion process $H(t)$ at singular boundary condition in case of $n = 1$ can be revealed and the changing pattern of the sample stability at the point of $H(t)$ can be identified. Hence, the influence of the stochastic perturbation on the homoclinic divergence can be understood.

In case of $n = 1$, $E_1(n)$ and $E_2(n)$ have the mathematical properties as:

$$\lim_{n \to 1} E_1(n) = +\infty$$

$$\lim_{n \to 1} E_2(n) = 1$$

So it can be deduced that if $n = 1$ then $\bar{m}(H)|_{n=1} = \bar{\sigma}^2(H)|_{n=1} = 0$, and $n = 1$ is the trap point of the diffusion process $H(t)$ and the singular boundary of the first kind.

Based on the definition of the diffusion exponent α_r, drift exponent β_r and character exponent c_r of the singular boundary of the first kind [28], α_r, β_r and c_r can be obtained:

$$\alpha_r = 2$$

$$\beta_r = 1$$

$$c_r = -\lim_{n \to 1} \frac{2\bar{m}(H)(1-n)^{\alpha_r - \beta_r}}{\bar{\sigma}^2(H)}$$

$$= -\lim_{n \to 1} \frac{\frac{4}{15}\left[\frac{C_1}{C_3(2-n)}\right]^3 \left[\frac{C_3\left(10x_0^2+5\sigma^2\right)}{2C_1} - 1\right](2-n)(1-n)^2}{\frac{4\sigma^2}{35}\left[\frac{C_1}{C_3(2-n)}\right]^4 (1-n)^2(n+16)}$$

$$c_r = -\frac{14C_3}{51C_1}\left[\frac{C_3\left(10x_0^2 + 5\sigma^2\right)}{2C_1} - 1\right] \tag{7.39}$$

With regard to the diffusion exponent α_r, drift exponent β_r and character exponent c_r of the singular boundary of the first kind, it is known that, if $c_r > \beta_r = 1$, then $n = 1$ is the repulsive natural boundary; if $c_r < \beta_r = 1$, then $n = 1$ is the attractive natural boundary; if $c_r = \beta_r = 1$, then $n = 1$ is the strictly natural boundary. So it is also known that $c_r = 1$ is a critical point at this point and the sample stability of $H(t)$ has a transition.

From Eq. 7.39, it is obtained that $c_r = 1$ pertains to $x_0 = \frac{C_1}{5C_3} - \frac{\sigma^2}{2} - 0.73\frac{C_1^2\sigma^2}{C_3^2}$. So one may get the analysis conclusions as follows.

(1) If $x_0 < \frac{C_1}{5C_3} - \frac{\sigma^2}{2} - 0.73\frac{C_1^2\sigma^2}{C_3^2}$, or $c_r > 1$, then the diffusion process $H(t)$ has a repulsive natural boundary at $n = 1$ $(H = 0)$, and here $H(t) = 0$ is the unsteady trivial solution.
(2) If $x_0 = \frac{C_1}{5C_3} - \frac{\sigma^2}{2} - 0.73\frac{C_1^2\sigma^2}{C_3^2}$, or $c_r = 1$, then $H(t)$ has a strict natural boundary at $n = 1$ $(H = 0)$, and $H(t) = 0$ also is the unsteady trivial solution.
(3) If $x_0 > \frac{C_1}{5C_3} - \frac{\sigma^2}{2} - 0.73\frac{C_1^2\sigma^2}{C_3^2}$, or $c_r < 1$, then $H(t)$ has an attractive natural boundary at $n = 1$ $(H = 0)$, and $H(t) = 0$ is the asymptotic steady trivial solution.

According to the above analysis, it is clear that while the divergence parameter x_0 reaches the critical value of $x_0 = \frac{C_1}{5C_3} - \frac{\sigma^2}{2} - 0.73\frac{C_1^2\sigma^2}{C_3^2}$, the sample steadiness of the trivial solution $H(t) = 0$ has a transition from unsteadiness to asymptotic steadiness; consequently, a subcritical stochastic bifurcation of stochastic differential kinetic system (7.4) will occur. The introduction of the white noise process leads to a drift of the divergence point of the non-linear stochastic differential kinetic system (7.4) in contrast to the homoclinic divergence of the non-linear deterministic differential kinetic system (7.4). Using the same method in the Sects. 7.3 and 7.4, the fact that stochastic bifurcations could still occur under some conditions indicates that such behaviour may be an important factor in the random behaviour of real crack-propagation.

7.6 Solution of Fatigue Damage FPK (Fokker-Planc-Kolgmorov) Equation

As is well known, deterministic fatigue damage evolution may be written into following general form as:

$$\frac{dx}{dt} = \varepsilon f(x, q_i) \tag{7.40}$$

where t is the time measuring, i.e., natural time or cycle number. q_i is the dimensionless variable. x is the state variable dependent on time t. From the definition and irreversibility of fatigue damage, it is possible to have $x \geq 0, f \geq 0$ and small dimensionless quantity $\varepsilon > 0$.

Randomizing Eq. 7.40 gives stochastic differential equation as:

$$\frac{dx}{dt} = \varepsilon f(x)h(t) \tag{7.41}$$

where $h(t)$ is the stochastic fluctuation of cyclic loading or microstructural behaviour of material independent on x.

Transforming the damage state space x in Eq. 7.41 into y:

$$y = \eta(x), \quad x = \eta^{-1}(y) \tag{7.42}$$

with:

$$f\left[\eta^{-1}(y)\right]\left|\frac{\partial y}{\partial x}\right|_{x=\eta^{-1}(y)} = 1 \tag{7.43}$$

then it is possible to have:

$$\frac{dy}{dt} = \varepsilon h(t) \tag{7.44}$$

where $\{y(t),\ t \in T\}$ is the Markov process associated with x and η is the monadic monotonic Borel function.

Then the drift coefficient $a(t)$ and diffusion coefficient $b(t)$ of the FPK equation corresponding to Eq. 7.44 are respectively:

$$a(t) = \varepsilon E[h(t)] \tag{7.45}$$

$$b(t) = 2\varepsilon^2\left\{E\left[h(t)\int_{-\infty}^{t} h(t_1)dt_1\right] - E[h(t)\cdot h(t)]\right\} \tag{7.46}$$

and the corresponding FPK equation of Eq. 7.44 is:

$$\frac{dp}{dt} = \mp a(t)\frac{\partial p(y)}{\partial y} + b(t)\frac{\partial^2 p(y)}{\partial y^2} \tag{7.47}$$

In the case of $[\partial y/\partial x]_{x=\eta^{-1}(x)} > 0$, then Eq. 7.47 takes minus, otherwise plus. Equation 7.47 demonstrates that $y(t)$ is the homogeneous space Markov process and transfer probability density function (PDF) $p(y)$ depends on t_0, t and $(y - y_0)$. Solving Eq. 7.47 yields:

$$p(y) = \left[2\pi\int_{t_0}^{t} b(t_1)dt_1\right]^{-\frac{1}{2}}\exp\left\{-\frac{\left[y - y_0 \pm \int_{t_0}^{t} a(t_1)dt_1\right]^2}{2\int_{t_0}^{t} b(t_1)dt_1}\right\} \tag{7.48}$$

From transformation equation (7.43), one has:

$$\frac{dy}{dx} = \pm\frac{1}{f(x)} \tag{7.49}$$

Integrating Eq. 7.49 gives:

$$y - y_0 = \pm\int_{x_0}^{x}\frac{dx}{f(x)} \tag{7.50}$$

From the inverse transformation of Eq. 7.43, transforming Eq. 7.48 and using Eq. 7.50, it is possible to have the solution of transfer PDF of fatigue damage Markov process as:

$$p(x) = \left[2\pi \int_{t_0}^{t} b(t_1)dt_1\right]^{-\frac{1}{2}} |f(x)|^{-1} \exp\left\{ -\frac{\left[\int_{x_0}^{x} \frac{dx}{f(x)} - \int_{t_0}^{t} a(t_1)dt_1\right]^2}{2\int_{t_0}^{t} b(t_1)dt_1} \right\} \tag{7.51}$$

If initial damage x_0 is a random variable with a PDF of $g(x_0)$ and the span of x_0 is $[0, x_c)$, here x_c is a deterministic damage upper limit, then the PDF of fatigue damage stochastic process *x*(*t*) is:

$$p(x) = \int_0^{x_c} p(x, x_0) g(x_0) dx_0 \tag{7.52}$$

7.7 Damage Probability Distributions for Fatigue Crack Formation and Propagation

According to the Miner rule, fatigue accumulation damage can be described with a scale function of *D*(*t*). Now, fatigue damage growth rate is:

$$\frac{dD}{dn_s} = \frac{1}{N_s} \tag{7.53}$$

where n_s is the fatigue loading cycle number. Assuming expected cycle number of fatigue stress in unit time is *n*(*t*), then:

$$dn_s = n(t) \cdot d(t) \tag{7.54}$$

From Eqs. 7.53 and 7.54, it can be shown that:

$$\frac{dD}{dt} = n(t)\frac{dD}{dn_s} = \frac{n(t)}{N_s} \tag{7.55}$$

where N_s is determined from generalized *S–N* surface (1.18). Substituting Eq. 1.18 into Eq. 7.55 yields:

$$\frac{dD}{dt} = \frac{n(t)}{C[s(t) - S_0]^m} \tag{7.56}$$

with:

$$s(t) = \frac{\sigma_b s_a(t)}{\sigma_b - s_m(t)}$$

where $s_a(t)$ and $s_m(t)$ are fatigue stress amplitude and mean respectively.

According to stochastic process theory [29], the FPK equation of Eq. 7.56 and the drift coefficient $a(t)$ and diffusion coefficient $b(t)$ become respectively:

$$\frac{dp}{dt} = -a(t)\frac{\partial p(D)}{\partial D} + b(t)\frac{\partial^2 p(D)}{\partial D^2} \tag{7.57}$$

$$a(t) = E\left[\frac{n(t)}{C[s(t) - S_0]^m}\right] \tag{7.58}$$

$$b(t) = 2\left\{E\left[\frac{n(t)}{C^2[s(t) - S_0]^m}\int_{-\infty}^{t}\frac{n(t_1)}{[s(t_1) - S_0]^m}dt_1\right] - E\left[\frac{n^2(t)}{C^2[s(t) - S_0]^{2m}}\right]\right\} \tag{7.59}$$

From Eq. 7.51, one has the solution of Eq. 7.57 satisfying the initial condition of $p(D, t_0 = 0) = \delta(D - D_0)$ as:

$$p(D, t|D_0, t_0) = \left[2\pi\int_{t_0}^{t} b(t_1)dt_1\right]^{\frac{1}{2}} \cdot \exp\left\{-\frac{(D - D_0) - \left[\int_{t_0}^{t} a(t_1)dt_1\right]^2}{2\int_{t_0}^{t} b(t_1)dt_1}\right\} \tag{7.60}$$

For a stationary stress process, $n(t) = n$=constant, and for a narrowband stress process, n can be taken as expected cycle number, thus fatigue damage growth rate can be written as:

$$\frac{dD}{dt} = \frac{n}{N_s} = \frac{n}{C[s(t) - S_0]^m} \tag{7.61}$$

Similarly, from Eq. 7.51, it is possible to have the transfer PDF of fatigue damage D satisfying initial condition as:

$$p(D, t|D_0, t_0) = \left[2\pi\int_{t_0}^{t} b(t_1)dt_1\right]^{\frac{1}{2}} n^{-1} \cdot \exp\left\{-\frac{\left[n^{-1}(D - D_0) - \left[\int_{t_0}^{t} a(t_1)dt_1\right]^2\right]}{2\int_{t_0}^{t} b(t_1)dt_1}\right\} \tag{7.62}$$

If $s(t)$ is the stationary process, then $a(D, t)$ and $b(D, t)$ are respectively:

$$a(D, t) = \mu, \quad b(D, t) = \sigma$$

From the PDF of fatigue damage D, the expected value of fatigue crack formation life is:

$$E[T] = \left\{-\int_{x_0}^{x} \exp\left[-\int_{x_0}^{x_2} \mu\sigma^{-1}dx_1\right]dx_2 + \int_{x_0}^{x_c} \exp\left[-\int_{x_0}^{x_2} \mu\sigma^{-1}dx_1\right]dx_2\right.$$

$$+\frac{\int_{x_0}^{x_c}\exp\left[-\int_{x_0}^{x_3}\mu\sigma^{-1}dx_1\right]\left\{\int_{x_0}^{x_3}\exp\left[\int_{x_0}^{x_2}\mu\sigma^{-1}dx_1\right]dx_2\right\}dx_3}{\int_{x_0}^{x_c}\exp\left[\int_{x_0}^{x_2}\mu\sigma^{-1}dx_1\right]dx_2}$$
$$\cdot\int_{x_0}^{x}\exp\left[\int_{x_0}^{x_2}\mu\sigma^{-1}dx_1\right]dx_2$$
$$-\int_{x_0}^{x}\exp\left[-\int_{x_0}^{x_3}\mu\sigma^{-1}dx_1\right]\left\{\int_{x_0}^{x_3}\exp\left[\int_{x_0}^{x_2}\mu\sigma^{-1}dx_1\right]dx_2\right\}dx_3$$
$$=\frac{\sigma^2}{\mu^2}\left\{\exp\left[\frac{\mu}{\sigma}(x-x_0)\right]-\exp\left[\frac{\mu}{\sigma}(x_c-x_0)\right]+\frac{\frac{\mu}{\sigma}(x_c-x_0)-1+\exp\left[-\frac{\mu}{\sigma}(x_c-x_0)\right]}{\exp\left[\frac{\mu}{\sigma}(x_c-x_0)\right]-1}\right\}$$
$$\cdot\left\{\exp\left[\frac{\mu}{\sigma}(x-x_0)\right]-1\right\}-\frac{\sigma}{\mu}(x-x_0)+\frac{\sigma^2}{\mu^2}-\frac{\sigma^2}{\mu^2}\exp\left[-\frac{\mu}{\sigma}(x-x_0)\right] \tag{7.63}$$

and second moment $E[T^2]$ is:

$$E\left[T^2\right]=\frac{\sigma}{\mu}\left\{-2\int_{x_0}^{x}E[T]\exp\left[-\frac{\mu}{\sigma}(x_2-x_0)\right]dx_2+2\int_{x_0}^{x_c}E[T]\exp\left[-\frac{\mu}{\sigma}(x_2-x_0)\right]dx_2\right.$$
$$\left.+\frac{2\int_{x_0}^{x_c}E[T]\exp\left[-\frac{\mu}{\sigma}(x_3-x_0)\right]\left\{\exp\left[\frac{\mu}{\sigma}(x_3-x_0)\right]-1\right\}dx_3}{\exp\left[\frac{\mu}{\sigma}(x_c-x_0)\right]-1}\right\}$$
$$\cdot\left\{\exp\left[\frac{\mu}{\sigma}(x-x_0)\right]-1\right\}$$
$$-2\frac{\sigma}{\mu}\int_{x_0}^{x}E[T]\exp\left[-\frac{\mu}{\sigma}(x_3-x_0)\right]\left\{\exp\left[\frac{\mu}{\sigma}(x_3-x_0)\right]-1\right\}dx_3 \tag{7.64}$$

Taking the plastic zone at fatigue crack tip into account and correcting for finite width, the fatigue crack growth rate formulation in Eq. 1.42 becomes

$$\frac{da}{dN}=C[\Delta X(s_{\max},s_{\min})]^{m_1}Y^{m_1}(a)(1-R)^{m_2} \tag{7.65}$$

Noting that stress ratio is defined as:

$$R=\frac{s_{\min}}{s_{\max}} \tag{7.66}$$

Substituting Eq. 7.66 into Eq. 7.65 yields:

$$\frac{da}{dN}=C[\Delta X(s_{\max},s_{\min})]^{m_1}Y^{m_1}(a)\left(\frac{s_{\max}-s_{\min}}{s_{\max}}\right)^{m_2} \tag{7.67}$$

In the case where the expected cycle number of fatigue stress in unit time is $n(t)$, then damage growth rate in fatigue crack growth is:

$$\frac{dD(t)}{dt} = Cn(t)[\Delta X(s_{\max}(t), s_{\min}(t))]^{m_1} Y^{m_1}(D)\left(\frac{s_{\max}(t) - s_{\min}(t)}{s_{\max}(t)}\right)^{m_2} \tag{7.68}$$

Letting $z(t) = \Delta X[s_{\max}(t), s_{\min}(t)]$, $s(t) = s_{\max}(t) - s_{\min}(t)$, then Eq. 7.68 becomes:

$$\frac{dD(t)}{dt} = Cn(t)z^{m_1}(t)Y^{m_1}(D)\left(\frac{s(t)}{s_{\max}(t)}\right)^{m_2} \tag{7.69}$$

The drift coefficient $a(D, t)$ and diffusion coefficient $b(D, t)$ of the FPK Equation pertinent to Eq. 7.69 are respectively:

$$a(D,t) = E\left[Cn(t)z^{m_1}(t)Y^{m_1}(D)\left(\frac{s(t)}{s_{\max}(t)}\right)^{m_2}\right] \tag{7.70}$$

$$\begin{aligned} b(D,t) = 2\Bigg\{ & E\left[C^2 n(t)z^{m_1}(t)Y^{m_1}(D)\left(\frac{s(t)}{s_{\max}(t)}\right)^{m_2} \right. \\ & \left. \cdot \int_{-\infty}^{t} n(t_1)z^{m_1}(t_1)Y^{m_1}(D)\left(\frac{s(t_1)}{s_{\max}(t_1)}\right)^{m_2} dt_1\right] \cdot \\ & - E\left[C^2 n^2(t)z^{2m_1}(t)Y^{2m_1}(D)\left(\frac{s(t)}{s_{\max}(t)}\right)^{2m_2}\right] \end{aligned}$$

Similarly, the solution of the FPK equation satisfying the initial condition of $p(D, t = t_0 = 0) = \delta(D - D_0)$ becomes:

$$\begin{aligned} p(D,t|D_0,t_0) = C & \left[2\pi \int_{t_0}^{t} b(t_1)dt_1\right]^{\frac{1}{2}} n(t)z^{m_1}(t)Y^{m_1}(D)\left(\frac{s(t)}{s_{\max}(t)}\right)^{m_2} \\ & \cdot \exp\left\{ - \frac{\left[\int_{D_0}^{D} Y^{m_1}(x_1)dx_1 - \int_{t_0}^{t} a(t_1)dt_1\right]}{2\int_{t_0}^{t} b(t_1)dt_1} \right\} \end{aligned} \tag{7.72}$$

Letting $F(t) = \frac{Cz^{m_1}(t)s^{m_2}(t)}{s_{\max}^{m_2}(t)}$, then $F(t)$ is the stochastic process. If $F(t)$ is a stationary process, then $a(D,t) = \mu Y^{m_1}(D)$ and $b(D,t) = \sigma Y^{2m_1}(D)$. From the PDF of fatigue damage D, the time expected value of fatigue crack growth first passing critical state is:

$$\begin{aligned} E[T] = \mu^2\sigma^{-2} & \left\{ - \int_{x_0}^{x} \exp\left[- \int_{x_0}^{x_2} Y^{m_1}(x_1)dx_1 \right] dx_2 \right\} \\ & + \int_{x_0}^{x_c} \exp\left[- \int_{x_0}^{x_2} Y^{m_1}(x_1)dx_1 \right] dx_2 \end{aligned}$$

$$+\frac{\int_{x_0}^{x_c}\exp\left[-\int_{x_0}^{x_3}Y^{m_1}(x_1)dx_1\right]\left\{\int_{x_0}^{x_3}\exp\left[\int_{x_0}^{x_2}Y^{m_1}(x_1)dx_1\right]dx_2\right\}dx_3}{\int_{x_0}^{x_c}\exp\left[\int_{x_0}^{x_2}Y^{m_1}(x_1)dx_1\right]dx_2}\Bigg\}$$
$$\cdot\int_{x_0}^{x}\exp\left[\int_{x_0}^{x_2}Y^{m_1}(x_1)dx_1\right]dx_2$$
$$-\mu^2\sigma^{-2}\int_{x_0}^{x}\exp\left[-\int_{x_0}^{x_3}Y^{m_1}(x_1)dx_1\right]\left\{\int_{x_0}^{x_3}\exp\left[\int_{x_0}^{x_2}Y^{m_1}(x_1)dx_1\right]dx_2\right\}dx_3 \tag{7.73}$$

and second moment $E[T^2]$ is:

$$E[T^2]=\mu^2\sigma^{-2}\left\{-2\int_{x_0}^{x}E[T]\exp\left[-\int_{x_0}^{x_2}Y^{m_1}(x_1)dx_1\right]dx_2\right\}$$
$$+2\int_{x_0}^{x_c}E[T^n]\exp\left[-\int_{x_0}^{x_2}Y^{m_1}(x_1)dx_1\right]dx_2$$
$$+2\frac{\int_{x_0}^{x_c}E[T]\exp\left[-\int_{x_0}^{x_3}Y^{m_1}(x_1)dx_1\right]\left\{\int_{x_0}^{x_3}\exp\left[\int_{x_0}^{x_2}Y^{m_1}(x_1)dx_1\right]dx_2\right\}dx_3}{\int_{x_0}^{x_c}\exp\left[\int_{x_0}^{x_2}Y^{m_1}(x_1)dx_1\right]dx_2}\Bigg\}$$
$$\cdot\int_{x_0}^{x}\exp\left[\int_{x_0}^{x_2}Y^{m_1}(x_1)dx_1\right]dx_2$$
$$-2\mu^2\sigma^{-2}\int_{x_0}^{x}E[T]\exp\left[-\int_{x_0}^{x_3}Y^{m_1}(x_1)dx_1\right]$$
$$\cdot\left\{\int_{x_0}^{x_3}\exp\left[\int_{x_0}^{x_2}Y^{m_1}(x_1)dx_1\right]dx_2\right\}dx_3 \tag{7.74}$$

References

1. Langford SC, Ma Z, Dickinson JT (1989) Photon-emission as a probe of chaotic processes accompanying fracture. J Mater Res 4(5):1272–1279
2. Ma ZY, Langford SC, Dickinson JT, Engelhard MH, Baer DR (1991) Fractal character of crack-propagation in epoxy and epoxy composites as revealed by photo-emission during fracture. J Mater Res 6(1):183–195
3. Mandelbrot BB, Passoja DE, Paullay AJ (1984) Fractal character of fracture surfaces of metals. Nature 308:721–722
4. Underwood EE, Banerji K (1986) Fractals in fractography. J Mater Sci Eng 80:1–14
5. Paude CS, Richards LE, Louat N, Dempsey BD, Schwoeble AJ (1987) Fractal characterization of fracture surface. Acta Metallurgica Mater 35:1633–1637
6. Mu ZQ, Lung CW (1988) Studies on the fractal dimension and fracture toughness of steel. J Phys D Appl Phys 21:848–850
7. Cahn RW (1989) Fractal dimension and fracture. Nature 388:201–202

8. Dauskardt RH, Hanbensak F, Ritchie RO (1990) On the interpretation of the fractal character of fracture surfaces. Acta Metallurgica Mater 38:143–159
9. Lei W, Chen B (1995) Fractal characterization of some fracture phenomena. Eng Fract Mech 50(2):149–155
10. Markworth AJ, Mccoy JK, Rollins RW (1988) Chaotic dynamics in an atomistic model of environmentally assisted fracture. J Mater Res 3(4):675–686
11. Mohan R, Markworth AJ, Rollins RW (1994) Effect of dissipation and driving on chaotic dynamics in an atomistic crack tip model. Model Simul Mater Sci Eng 2:659–676
12. Tan HL, Yang W (1995) Nonlinear motion of atoms at a crack tip during dislocation emission process. J Appl Phys 78:7026–7034
13. Tan HL, Yang W (1996) Nonlinear motion of crack tip atoms during cleavage process. Int J Fract 77(3):199–212
14. Yang W, Tan HL (1996) Chaotic atom motion excited by fracture. Mater Sci Res Int 2(1): 1–12
15. Xiong J, Gao Z, Liu X, Sun XF (2000) On stochastic bifurcation behaviour of the atom motion at the tip of crack in fatigue damage system. Acta Phys Sin 49:49–53
16. Xiong JJ (2006) A nonlinear fracture differential kinetic model to depict chaotic atom motions at a fatigue crack tip based on the differentiable manifold methodology. Chaos Solitons Fractals 29(5):1240–1255
17. Rice JR (1992) Dislocation nucleation from a crack tip-an analysis based on the peierls concept. J Mech Phys Solids 40(2):239–271
18. Cherepanov GP (1974) Mechanics of brittle fracture. Nauka Press, Moscow
19. Freund LB (1990) Dynamic fracture mechanics. Cambridge University Press, Cambridge
20. Rutzel S, Lee SI, Raman A (2003) Nonlinear dynamics of atomic-force-microscope probe driven in Leonard-Jones potential. Proc R Soc Lond (series A) 459(2036):1925–1948
21. Cheung KS, Yip S (1994) A molecular-dynamics simulation of crack-tip extension: The brittle-to-ductile transition. Model Simul Mater Sci Eng 2:865–892
22. Hu Z, Fukuyama S, Yokogawa K, Okamoto S (1999) Hydrogen embrittlement of a single crystal of iron on a nanometre scale at a crack tip by molecular dynamics. Model Simul Mater Sci Eng 7:541–551
23. Belytschko T, Xiao SP, Schatz GC, Ruoff RS (2002) Atomistic simulations of nanotube fracture. Phys Rev B 65(23):235–430
24. Kapitaniak T (1998) Chaos for engineers. Theory, applications and control. Springer-Verlag, Berlin
25. Guckenheimer J, Holmes P (1983) Nonlinear oscillations, dynamical systems, and bifurcations of vector fields. Springer-Verlag, New York
26. Ito K, Mckean HP Jr (1965) Diffusion processes and their sample paths. Springer-Verlag, New York
27. Khasminskii RZ (1967) Necessary and sufficient condition for the asymptotic stability of linear stochastic system. Theory Probab Appl 12:144–147
28. Zhang ZY and Kozin F (1990) New developments in almost sure sample stability of nonlinear stochastic dynamic systems. In: Proceeding of IEEE international conference on system engineering, Pittsburg, 1990, pp 117–122
29. Zhu WQ (1992) Random oscillations. Science Press, Beijing